T0224630

SpringerBriefs in Physics

Albert Petrov · Jose Roberto Nascimento ·
Paulo Porfirio

Introduction to Modified Gravity

Second Edition

 Springer

Albert Petrov ⓘD
Departamento de Fisica
Federal University of Paraíba
João Pessoa, Paraíba, Brazil

Jose Roberto Nascimento ⓘD
Departamento de Fisica
Federal University of Paraíba
João Pessoa, Paraíba, Brazil

Paulo Porfirio ⓘD
Departamento de Fisica
Federal University of Paraíba
João Pessoa, Paraíba, Brazil

ISSN 2191-5423   ISSN 2191-5431 (electronic)
SpringerBriefs in Physics
ISBN 978-3-031-46633-5   ISBN 978-3-031-46634-2 (eBook)
https://doi.org/10.1007/978-3-031-46634-2

This Springer imprint is published by the registered company Springer Nature Switzerland AG
The registered company address is: Gewerbestrasse 11, 6330 Cham, Switzerland

Paper in this product is recyclable.

Preface

In this book, we review various modifications of the Einstein gravity, First, we consider theories where only the purely geometric sector is changed. Second, we review scalar-tensor gravities. Third, we examine vector-tensor gravity models and the problem of Lorentz symmetry breaking in a curved space-time. Fourth, we present some results for the Horava-Lifshitz gravity. Fifth, we consider nonlocal extensions for gravity. Also, we give some comments on non-Riemannian gravity theories. We close the book with the discussion of perspectives of modified gravity.

The authors are grateful to Profs. B. Altschul, M. Gomes, T. Mariz, G. Olmo, E. Passos, M. Rebouças, A. F. Santos, A. J. da Silva, and J. B. Fonseca-Neto (in memoriam), for fruitful collaboration and interesting discussions. The work has been partially supported by CNPq.

João Pessoa, Brazil

Albert Petrov
Jose Roberto Nascimento
Paulo Porfirio

Contents

Chapter 1
Einstein Gravity and the Need for Its Modification

General relativity (GR) is clearly one of the most successful physical theories. Being formulated as a natural development of special relativity, it has made a number of fundamental physical predictions which have been confirmed experimentally with a very high degree of precision. Among these predictions, a special role is played by expansion of the Universe and precession of Mercure perihelion, which have been proved many years ago, while other important claims of GR, such as gravitational waves and black holes, have been confirmed through direct observations only recently (an excellent review of various experimental tests of gravity can be found in [1]).

By its concept, GR is an essentially geometric theory. Its key idea consists in the fact that the gravitational field manifests itself through modifications of the space-time geometry. Thus, one can develop a general theory of gravity where the fields characterizing geometry, that is, metric and connection, become dynamical variables so that a nontrivial space can be described in terms of curvature, torsion and non-metricity. It has been argued in [2] that there are eight types of geometry characterized by possibilities of zero or non-zero curvature tensor, torsion and so-called homothetic curvature tensor, with all these objects constructed on the base of metric and connection. Nevertheless, the most used formulation of gravity is based on the (pseudo-)Riemannian approach where the connection is symmetric and completely characterized by the metric. Within this book, we concentrate namely on the (pseudo-)Riemannian description of gravity where the action is characterized by functions of geometric invariants constructed on the basis of the metric (i.e. various contractions of Riemann curvature tensor, its covariant derivatives and a metric), and possibly some extra fields, scalar or vector ones, and only in Chap. 7 we discuss theories of gravity defined on a non-Riemannian manifold. So, let us introduce some basic definitions of quantities used within the Riemannian approach.

By definition, the infinitesimal line element in curved spaces is defined as $ds^2 = g_{\mu\nu}(x)dx^\mu dx^\nu$. The metric tensor $g_{\mu\nu}(x)$ is considered as the only independent dynamical variable in our theory. As usual, the action must be a (Riemannian) scalar, and for the first step, it is assumed to involve no more than the second derivatives of the metric tensor, in the whole analogy with other field theory models where the action involves only up to the second derivatives. The unique scalar involving

© The Author(s), under exclusive license to Springer Nature Switzerland AG 2023
A. Petrov et al., *Introduction to Modified Gravity*,
SpringerBriefs in Physics, https://doi.org/10.1007/978-3-031-46634-2_1

only second derivatives is a scalar curvature R (throughout the book, we follow the definitions and conventions from the book [3] except for special cases):

$$R = g^{\mu\nu} R_{\mu\nu}; \quad R_{\mu\nu} = R^{\alpha}{}_{\mu\alpha\nu};$$
$$R^{\kappa}{}_{\lambda\mu\nu} = \partial_{\mu}\Gamma^{\kappa}_{\lambda\nu} - \partial_{\mu}\Gamma^{\kappa}_{\lambda\nu} + \Gamma^{\kappa}_{\rho\mu}\Gamma^{\rho}_{\lambda\nu} - \Gamma^{\kappa}_{\rho\nu}\Gamma^{\rho}_{\lambda\mu}, \tag{1.1}$$

where $\Gamma^{\mu}_{\nu\lambda}$ are the Christoffel symbols, that is, affine connections expressed in terms of the metric tensor as

$$\Gamma^{\mu}_{\nu\lambda} = \frac{1}{2}g^{\mu\rho}(\partial_{\nu}g_{\rho\lambda} + \partial_{\lambda}g_{\rho\nu} - \partial_{\rho}g_{\nu\lambda}). \tag{1.2}$$

The Einstein–Hilbert action is obtained as an integral from the scalar curvature over the D-dimensional space-time:

$$S = \int d^{D}x\sqrt{|g|}(\frac{1}{2\kappa^2}R + \mathcal{L}_m), \tag{1.3}$$

where g is the determinant of the metric. We assume the signature to be $(+ - --)$. The $\kappa^2 = 8\pi G$ is the gravitational constant (it is important to note that its mass dimension in D-dimensional space-time is equal to $2 - D$, but within this book we concentrate on the usual case $D = 4$); nevertheless, in some cases we will define it to be equal to 1. The $\mathcal{L}_m$ is the matter Lagrangian.

Varying the action with respect to the metric tensor, we obtain the Einstein equations:

$$G_{\mu\nu} \equiv R_{\mu\nu} - \frac{1}{2}Rg_{\mu\nu} = \kappa^2 T_{\mu\nu}, \tag{1.4}$$

where $T_{\mu\nu}$ is the energy-momentum tensor of the matter. The conservation of the energy-momentum tensor presented by the condition $\nabla_{\mu}T^{\mu\nu} = 0$ is clearly consistent with the Bianchi identities $\nabla_{\mu}G^{\mu\nu} = 0$.

One should emphasize several most important solutions of these equations for the four-dimensional space-time. The first one is the Schwarzschild metric, which solves the vacuum Einstein equations, $T_{\mu\nu} = 0$, and describes the simplest black hole with mass M. The corresponding space-time line element looks like

$$ds^2 = (1 - \frac{2M}{r})c^2dt^2 - (1 - \frac{2M}{r})^{-1}dr^2 - r^2(d\theta^2 + \sin^2\theta d\phi^2). \tag{1.5}$$

Actually, this metric is a particular case of the more generic static spherically symmetric metric (SSSM).

The second one is the Friedmann-Robertson-Walker (FRW) metric describing the simplest (homogeneous and isotropic) cosmological solution whose line element is

$$ds^2 = c^2 dt^2 - a^2(t) \left(\frac{dr^2}{1 - kr^2} + r^2(d\theta^2 + \sin^2 \theta d\phi^2) \right), \tag{1.6}$$

where $a(t)$ is the scale factor, and $k = 1, 0, -1$ for positive, zero and negative curvature, respectively. The matter, in this case, is given by the relativistic perfect fluid:

$$\kappa^2 T_{\mu\nu} = (\rho + p)v_\mu v_\nu + p g_{\mu\nu}, \tag{1.7}$$

where ρ is the matter density, v^μ is the 4-velocity of a point-particle of the matter and p is its pressure; in many cases, one employs the equation of state $p = \omega\rho$, with ω being a constant characterizing the kind of the matter.

Besides these solutions, an important example is represented also by the Gödel metric with the line element [4]:

$$ds^2 = a^2[(dt + e^x dy)^2 - dx^2 - \frac{1}{2}e^{2x}dy^2 - dz^2], \tag{1.8}$$

which, just as the FRW metric, solves Einstein equations supported by the fluid-like matter source, i.e.

$$\kappa^2 T_{\mu\nu} = \kappa^2 \rho v_\mu v_\nu + \Lambda g_{\mu\nu}, \tag{1.9}$$

but in this case one has $v^\mu = \frac{1}{a}$, $\kappa^2 \rho = \frac{1}{a^2}$ and $\Lambda = -\frac{1}{2a^2}$. Namely, these solutions and their direct generalizations will be considered within our book.

An important generalization of the Gödel metric is the (cylindrically symmetric) class of Gödel-type metrics defined in [5]:

$$ds^2 = (dt + H(r)d\phi)^2 - D^2(r)d\phi^2 - dr^2 - dz^2, \tag{1.10}$$

where the following conditions of space-time homogeneity are assumed:

$$\frac{H'}{D} = 2\Omega, \quad \frac{D''}{D} = m^2, \tag{1.11}$$

with Ω and m being constants. This metric will be considered within our book in different modified gravity models. The interesting special case of this metric corresponding to $m^2 = 4\Omega^2$, discussed in great detail in [5], is called the Rebouças-Tiomno (RT) metric. Following the methodology described in [5], we consider three cases of H and D consistent with the conditions of space-time homogeneity of the metric (1.11):
(i) hyperbolic, $H = \frac{2\Omega}{m^2}[\cosh mr - 1]$, $D = \frac{1}{m}\sinh mr$;
(ii) trigonometric, $H = \frac{2\Omega}{\mu^2}[1 - \cos \mu r]$, $D = \frac{1}{m}\sin \mu r$; $\mu^2 = -m^2$;
(iii) linear, $H = \Omega r^2$, $D = r$.
Repeating the argumentation from [5], one immediately sees that for $0 < m^2 < 4\Omega^2$, there is a non-causal region with $r > r_c$, where $\sinh^2 \frac{mr_c}{2} = (\frac{4\Omega^2}{m^2} - 1)$, while at $m^2 \geq$

$4\Omega^2$, causality violation does not occur, and at $m^2 = -\mu^2 < 0$, there is an infinite sequence of causal and non-causal regions.

Now, let us make an introduction to quantum gravity. Indeed, it is natural to expect that gravity, in a whole analogy with electrodynamics and other field theories, must be quantized. To do it, one can follow the approach developed by 't Hooft and Veltman [6, 7]. We start with splitting of the dynamic metric $g_{\mu\nu}$ into a sum of the background part $\bar{g}_{\mu\nu}$ and the quantum fluctuation $h_{\mu\nu}$:

$$g_{\mu\nu} = \bar{g}_{\mu\nu} + \kappa h_{\mu\nu}, \tag{1.12}$$

where the κ is introduced to change the mass dimension of $h_{\mu\nu}$ to 1. As a result, the action can be expanded in infinite power series in $h_{\mu\nu}$. For the first step, we can choose the flat background, i.e. $\bar{g}_{\mu\nu} = \eta_{\mu\nu}$. The lowest, quadratic contribution to the Lagrangian of $h_{\mu\nu}$ is

$$\mathcal{L}_0 = \frac{1}{4}\partial_\mu h^\alpha_\alpha \partial^\mu h^\beta_\beta - \frac{1}{2}\partial_\beta h^\alpha_\alpha \partial^\mu h^\beta_\mu - \frac{1}{4}\partial_\mu h_{\alpha\beta}\partial^\mu h^{\alpha\beta} + \frac{1}{2}\partial_\alpha h_{\nu\beta}\partial^\nu h^{\alpha\beta}, \tag{1.13}$$

where the indices of $h_{\alpha\beta}$ are raised and lowered with the flat Minkowski metric. The Lagrangian (1.13) is called the Fierz-Pauli Lagrangian; it is used in constructing some generalizations of gravity.

The corresponding (second-order) equations of motion are actually the linearized Einstein equations:

$$G^{(0)}_{\mu\nu} \equiv -\frac{1}{2}(\partial^\lambda \partial_\mu h_{\lambda\nu} + \partial^\lambda \partial_\nu h_{\lambda\mu}) + \frac{1}{2}\Box h_{\mu\nu} + \frac{1}{2}\eta_{\mu\nu}\partial_\alpha\partial_\beta h^{\alpha\beta} -$$
$$- \frac{1}{2}\eta_{\mu\nu}\Box h^\lambda_\lambda + \frac{1}{2}\partial_\mu\partial_\nu h^\lambda_\lambda = 0. \tag{1.14}$$

The action is invariant with respect to linearized gauge transformations $\delta h_{\mu\nu} = \partial_\mu\xi_\nu + \partial_\nu\xi_\mu$; also, the linearized Bianchi identities $\partial^\mu G^{(0)}_{\mu\nu} = 0$ take place. As a consequence, afterwards one must fix the gauge, which can be done by adding the term

$$\mathcal{L}_{GF} = -\frac{1}{2}C_\mu C^\mu, \tag{1.15}$$

where $C_\mu = \partial^\alpha h_{\alpha\mu} - \frac{1}{2}\partial_\mu h^\alpha_\alpha$, so one has a new Lagrangian

$$\mathcal{L} = \mathcal{L}_0 - \frac{1}{2}C_\mu C^\mu = -\frac{1}{4}\partial_\mu h_{\alpha\beta}\partial^\mu h^{\alpha\beta} + \frac{1}{8}\partial_\mu h^\alpha_\alpha \partial^\mu h^\beta_\beta, \tag{1.16}$$

which can be rewritten as

$$\mathcal{L} = -\frac{1}{2}\partial^\lambda h_{\alpha\beta} V^{\alpha\beta\mu\nu}\partial_\lambda h_{\mu\nu}, \tag{1.17}$$

where $V^{\alpha\beta\mu\nu} = \frac{1}{2}\eta^{\alpha\mu}\eta^{\beta\nu} - \frac{1}{4}\eta^{\alpha\beta}\eta^{\mu\nu}$, which implies the following propagator in the momentum space:

$$< h_{\alpha\beta}(-k)h_{\mu\nu}(k) >= i\frac{\eta_{\mu\alpha}\eta_{\nu\beta} + \eta_{\nu\alpha}\eta_{\mu\beta} - \frac{2}{D-2}\eta_{\mu\nu}\eta_{\alpha\beta}}{k^2 - i\epsilon}, \qquad (1.18)$$

where D is the space-time dimension (the singularity at $D = 2$ reflects the fact that the $D = 2$ Einstein–Hilbert action is a pure surface term).

Now, let us expand the Einstein–Hilbert action (1.3) in power series in $h_{\mu\nu}$ by making again the substitution (1.12) but with the arbitrary background $\bar{g}_{\mu\nu}$. In this case, the metric determinant and curvature scalar are expanded up to the second order in h as (cf. [6, 7])

$$\sqrt{|g|} \rightarrow \sqrt{|\bar{g}|}(1 + \frac{1}{2}h_\alpha^\alpha - \frac{1}{4}h_\alpha^\beta h_\beta^\alpha + \frac{1}{8}(h_\alpha^\alpha)^2 + \dots); \qquad (1.19)$$

$$R \rightarrow \bar{R} + \Box h_\beta^\beta - \nabla^\alpha\nabla^\beta h_{\alpha\beta} - \bar{R}^{\alpha\beta}h_{\alpha\beta} - \frac{1}{2}\nabla_\alpha(h_\mu^\beta h_\beta^{\mu,\alpha}) + \frac{1}{2}\nabla_\beta[h_\nu^\beta(2h_{,\alpha}^{\nu\alpha} - h_\alpha^{\alpha,\nu})] +$$

$$+ \frac{1}{4}(h_{\beta,\alpha}^\nu + h_{\alpha,\beta}^\nu - h_{\alpha\beta}^{,\nu})(h_\nu^{\beta,\alpha} + h_{,\nu}^{\beta\alpha} - h_\nu^{\alpha,\beta}) -$$

$$- \frac{1}{4}(2h_{,\alpha}^{\nu\alpha} - h_\alpha^{\alpha,\nu})h_{\beta,\nu}^\beta - \frac{1}{2}h^{\nu\alpha}h_{\beta,\nu\alpha}^\beta + \frac{1}{2}h_\alpha^\nu\nabla_\beta(h_\nu^{\beta,\alpha} + h_{,\nu}^{\beta\alpha} - h_\nu^{\alpha,\beta}) + h_\beta^\nu h_\alpha^\beta\bar{R}_\nu^\alpha,$$

where $h_{\beta,\alpha}^\mu \equiv \nabla_\alpha h_\beta^\mu$, etc. and the covariant derivative, Ricci scalar and tensor and other objects with the bar above are constructed on the base of the background metric.

Also, we must introduce the Faddeev-Popov (FP) ghosts whose Lagrangian looks like [6]

$$\mathcal{L}_{gh} = \bar{C}^\mu(\Box C_\mu - \bar{R}_{\mu\nu}C^\nu), \qquad (1.20)$$

where C^μ, $\bar{C}^\mu$ are the ghosts. The expansion of the curvature scalar given by (1.19) together with the ghost action (1.20) is sufficient for the one-loop calculations. If one adds the coupling of the gravity to the scalar matter ϕ (which is further integrated out), through the Lagrangian $\mathcal{L}_{sc} = -\frac{1}{2}\sqrt{|g|}g^{\mu\nu}\partial_\mu\phi\partial_\nu\phi$, the following paradigmatic result for the one-loop counterterm arising in the purely gravitational sector [6, 7] can be found, within the dimensional regularization in D-dimensional space-time:

$$\mathcal{L}^{(1)} = \frac{\sqrt{|g|}}{8\pi^2(D-4)}\left(\frac{1}{120}R^2 + \frac{7}{20}R_{\mu\nu}R^{\mu\nu}\right). \qquad (1.21)$$

This result will be used in the next chapter to formulate the simplest renormalizable gravity model.

Returning to the classical aspects of gravity, we note again that many predictions of GR were confirmed experimentally, including, on the one side, the expansion of the Universe (which is discussed now in any textbook on GR, for example, in [3]), and, on the another side, the existence of gravitational waves whose detection within the LIGO-Virgo experiments was reported in [8]. In this context, it is worth to mention also obtaining the image of the shadow of a black hole with the use of the Event Horizon Telescope (EHT) observations [9], which played a fundamental role within the studies of black holes. Nevertheless, it turns out that there are problems which cannot be solved by GR itself, so, the theory of gravity requires some modifications. Actually, there are two most important difficulties which GR faced. The first one is related with the quantum description of gravity—indeed, the gravitational constant κ^2 has a negative mass dimension, precisely equal to $2 - D$ in D-dimensional space-time; thus, the four-dimensional Einstein–Hilbert gravity is non-renormalizable, i.e. its consistent description must involve an infinite number of counterterms (an excellent review on quantum calculations in gravity is presented in the book [10]). The second difficulty consists in the fact that the cosmic acceleration whose discovery was reported in [11] has not been predicted theoretically since it does not admit explanations within GR.

Therefore, the problem of possible modifications of gravity arises naturally. Actually, although the first attempts to introduce modified gravity have been carried out much earlier, the detection of gravitational waves [8] and obtaining the image of a black hole [9] strongly increased attention to modified gravity models.

The simplest attempt to solve the cosmic acceleration problem is based on introducing the cosmological constant Λ, i.e. we add to the action (1.3) the extra term $S_\Lambda = \frac{1}{\kappa^2} \Lambda \int d^4x \sqrt{|g|}$, so, in the l.h.s. of (1.4), the additive term $-\Lambda g_{\mu\nu}$ (or, as is the same, the extra term $\Lambda g_{\mu\nu}$ in the r.h.s.) will arise. It is easy to see that for the FRW metric, the components of the Ricci tensor and the scalar curvature in spherical coordinates, at $c = 1$, are

$$R_{00} = \frac{3\ddot{a}}{a}; \quad R_{11} = -\frac{a\ddot{a} + 2\dot{a}^2 + 2k}{1 - kr^2};$$
$$R_{22} = -r^2(a\ddot{a} + 2\dot{a}^2 + 2k); \quad R_{33} = -r^2 \sin^2\theta(a\ddot{a} + 2\dot{a}^2 + 2k);$$
$$R = 6\left(\frac{\ddot{a}}{a} + \frac{\dot{a}^2}{a^2} + \frac{k}{a^2}\right). \tag{1.22}$$

In this scenario, the (00) component of Einstein equations, together with the equation obtained from the combination of (ii) and (00) components, with $c = 1$ and $\kappa^2 = 8\pi G$, yield

$$\frac{\dot{a}^2 + k}{a^2} = \frac{1}{3}(8\pi G\rho + \Lambda); \tag{1.23}$$
$$\frac{\ddot{a}}{a} = -\frac{4}{3}\pi G(\rho + 3p) + \frac{\Lambda}{3},$$

where $k = +1, 0, -1$ for positive, zero and negative scalar curvature. As is well known, originally Λ was introduced by Einstein in order to provide a static solution, but further de Sitter proved that the empty space with positive Λ will expand exponentially (indeed, in this case, for $k = 0$, one easily finds $a(t) = a(0)e^{\sqrt{\frac{\Lambda}{3}}t}$, and the

Hubble parameter $H = \frac{\dot{a}}{a}$ is a constant). As a consequence, after the discovery of the cosmic acceleration, the idea of the cosmological constant has been revitalized [12]. However, the cosmological constant, by astronomical observations, should be extremely small (about 120 orders less than a natural scale for it given by M_{Planck}^4, where $M_{Planck} = 1/\kappa$, in gravitational units, is the Planck mass), and this fact has no theoretical reason (the search for this explanation constitutes the famous cosmological constant problem [12]). Besides, the cosmological constant does not solve the problem of renormalizability of gravity.

There are two manners on how to extend the gravity in order to solve these problems. Within the first approach, we modify the Einstein–Hilbert action by introducing additive terms. Within the second approach, we suggest that the full description of gravity involves, besides the metric field, also some extra scalar or vector fields which must not be confused with matter being treated as ingredients of the gravity itself, so that usual results of GR are recovered, for example, when these fields are constant (the typical example is the Brans-Dicke gravity which we discuss further). In this review, we give a description of these approaches. It should be noted that among these approaches, an important role is played by adding new geometrical terms (and/or fields) aimed either to break the Lorentz/CPT symmetry or to introduce a supersymmetric extension of gravity. Within this review we also discuss these approaches.

The structure of this book is as follows. In Chap. 2, we present various models obtained through the modifications of the purely geometric sector. In Chap. 3, we consider various scalar-tensor gravity models, such as Chern-Simons and Brans-Dicke gravities, and galileons. In Chap. 4, we discuss vector-tensor gravity models and the problem of Lorentz symmetry breaking in gravity. In Chap. 5, we review the most interesting results in Horava-Lifshitz gravity. In Chap. 6, we present some results for nonlocal gravity. In Chap. 7, we give some general comments on the non-Riemannian geometry and then address two particular modified theories of gravity on non-Riemannian manifolds. Chapter 8 represents conclusions of our book.

Chapter 2
Modifications of the Purely Geometric Sector

2.1 Motivations

As we already noted in the Introduction, one of the ways to modify gravity consists in introducing additional terms to the geometric sector. Such terms are given by scalars constructed on the base of the metric tensor, i.e. these scalars are functions of the Riemann tensor, the Ricci tensor, possibly, their covariant derivatives, and the scalar curvature. In the simplest case the Lagrangian is the function of the scalar curvature only, so, the action is

$$S = \frac{1}{16\pi G} \int d^4x \sqrt{|g|} f(R), \tag{2.1}$$

where, $f(R)$ is a some function of the scalar curvature. Since the Einstein gravity is very well observationally confirmed, and the curvature of the Universe is known to be small, it is natural to suggest that $f(R) = R + \gamma R^n$, with $n \geq 2$, where γ is a small constant parameter, so that Einstein-Hilbert term dominates. The case $n = 2$ is very interesting by various reasons, from power-counting renormalizability to the possibility of cosmic acceleration, so, it will be discussed in details. However, other values of n, including even negative ones which called attention recently, are also interesting. Another generalization of this action is the suggestion that the Lagrangian depends also on invariants $Q = R_{\mu\nu} R^{\mu\nu}$ and $P = R_{\mu\nu\alpha\beta} R^{\mu\nu\alpha\beta}$, such class of theories is called $f(R, Q, P)$ gravity, the paradigmatic example of such theories is the Weyl gravity (see f.e. [13, 14] and references therein), where the Lagrangian includes the square of the Weyl tensor, in D dimensions given by

$$C_{\mu\nu\lambda\rho} = R_{\mu\nu\lambda\rho} - \frac{1}{D-2}(g_{\mu\rho} R_{\nu\lambda} + g_{\nu\lambda} R_{\mu\rho} - g_{\nu\rho} R_{\mu\lambda} - g_{\mu\lambda} R_{\nu\rho})$$

$$- \frac{1}{(D-1)(D-2)} R(g_{\mu\lambda} g_{\nu\rho} - g_{\nu\lambda} r_{\mu\rho}), \tag{2.2}$$

© The Author(s), under exclusive license to Springer Nature Switzerland AG 2023
A. Petrov et al., *Introduction to Modified Gravity*,
SpringerBriefs in Physics, https://doi.org/10.1007/978-3-031-46634-2_2

satisfying the zero trace condition $C^{\mu}_{\nu\mu\rho} = 0$. Besides of these situations, it is interesting also to abandon the restriction for the space-time to be four-dimensional. In this context we will consider also higher-dimensional space-times and discuss Lovelock gravities whose action involves higher curvature invariants.

2.2 R^2-Gravity

Let us start with the action

$$S = \frac{1}{2\kappa^2} \int d^4x \sqrt{|g|}(R + \alpha R_{\mu\nu}R^{\mu\nu} - \beta R^2) + S_{mat}. \tag{2.3}$$

A simple comparison of this expression with (1.19) shows that this action is of the second order in curvatures, i.e. of fourth order in derivatives, therefore the theory described by this action is called R^2-gravity. In principle, one can add also the square of the Riemann tensor, however, since in the four-dimensional space-time the Gauss-Bonnet term $\mathcal{G} = R^2 - 4R_{\mu\nu}R^{\mu\nu} + R_{\mu\nu\lambda\rho}R^{\mu\nu\lambda\rho}$ is a total derivative, the square of the Riemann tensor in $D = 4$ is not independent.

We see that the additive term in this action exactly matches the structure of the one-loop divergence arising in the pure Einstein gravity (1.21). Therefore, the theory (2.3) is one-loop renormalizable. Moreover, it is not difficult to show that no other divergences arise in the theory. Here, we demonstrate it in the manner similar to that one used within the background field method for the super-Yang-Mills theory [15]. Indeed, the propagator in this theory behaves as k^{-4}. Any vertex involves no more than four derivatives. Integration over internal momentum in any loop yields the factor 4, hence formally the superficial degree of divergence must be $\omega = 4L - 4P + 4V = 4$. However, we should take into account that this is the upper limit for ω, and each derivative acting to the external legs instead of the propagator decreases ω by 1. Since $R_{\mu\nu\lambda\rho}$, as well as the Ricci tensor, involves second derivatives, each external $R_{\mu\nu\lambda\rho}$, $R_{\mu\nu}$, R decreases the ω by 2. Hence, the R^2 or $R_{\mu\nu}R^{\mu\nu}$ contributions will display only logarithmic divergences, and higher-order contributions like R^3 will yield $\omega < 0$ being thus superficially finite. The presence of Faddeev-Popov (FP) ghosts evidently does not jeopardize this conclusion.

Let us discuss various aspects of the theory (2.3). We follow the argumentation presented in [16, 17]. First, one can write down the equations of motion:

$$H_{\mu\nu} \equiv (\alpha - 2\beta)\nabla_{\mu}\nabla_{\nu}R - \alpha\Box R_{\mu\nu} - (\frac{\alpha}{2} - 2\beta)g_{\mu\nu}\Box R +$$

$$+ 2\alpha R^{\rho\lambda}R_{\mu\rho\nu\lambda} - 2\beta R R_{\mu\nu} - \frac{1}{2}g_{\mu\nu}(\alpha R^{\rho\lambda}R_{\rho\lambda} - \beta R^2) +$$

$$+ \frac{1}{G}(R_{\mu\nu} - \frac{1}{2}Rg_{\mu\nu}) = T_{\mu\nu}. \tag{2.4}$$

Using these equations, one can find the Newtonian static limit of the theory. Proceeding in the same way as in GR, we can show that the gravitational potential in the non-relativistic limit is

$$\phi = h^{00} = \frac{1}{r} - \frac{4}{3}\frac{e^{-m_2 r}}{r} + \frac{1}{3}\frac{e^{-m_0 r}}{r}, \tag{2.5}$$

where $m_0 = (16\pi G\alpha)^{-1/2}$, and $m_2 = (32\pi G(3\beta - \alpha))^{-1/2}$. So, we find that the R^2-gravity involves massive modes displaying Yukawa-like contributions to the potential. Following the estimations from [16], the $m_{0,2}$ are about $10^{-17} M_{Pl}$. We note that the Birkhoff theorem is no more valid in this theory since there are mass-like parameters m_0, m_2, and instead of the Bianchi identities one will have $\nabla_\mu H^{\mu\nu} = 0$.

Then, it is interesting to discuss cosmological solutions in this theory. A remarkable feature of the R^2-gravity consists in the fact that it was the first gravity model to predict accelerated expansion of the Universe much before its observational discovery. The pioneer role was played by the paper [18], where terms of higher orders in curvature generated by some anomaly have been introduced to the equation of motion, so the resulting equation, for the vacuum, looks like

$$G_{\mu\nu} = k_1(R_\mu^\lambda R_{\nu\lambda} - \frac{2}{3}RR_{\mu\nu} - \frac{1}{2}g_{\mu\nu}R_{\alpha\beta}R^{\alpha\beta} + \frac{1}{4}g_{\mu\nu}R^2) +$$
$$+ k_2(\nabla_\nu\nabla_\mu R - 2g_{\mu\nu}\Box R - 2RR_{\mu\nu} + \frac{1}{2}g_{\mu\nu}R^2), \tag{2.6}$$

where k_1, k_2 are constants. Many terms in the r.h.s. of this equation are present also in (2.4), actually, at $\alpha = 0$ and $k_1 = 0$ these equations coincide up to some numerical coefficients, so, their solutions are not very different. Substituting the FRW metric into (2.6), we arrive at

$$\frac{\dot{a}^2 + k}{a^2} = \frac{1}{H^2}\left(\frac{\dot{a}^2 + k}{a^2}\right)^2 - \tag{2.7}$$
$$- \frac{1}{M^2}\left(\frac{\dot{a}}{a^2}\frac{d^3 a}{dt^3} - \frac{\ddot{a}^2}{a^2} + 2\frac{\ddot{a}\dot{a}^2}{a^3} - 3(\frac{\dot{a}}{a})^4 - 2k\frac{\dot{a}^2}{a^4} + \frac{k^2}{a^4}\right),$$

where $H^2 = \frac{\pi}{8Gk_1}$, $M^2 = -\frac{\pi}{8Gk_2}$, with $k_2 < 0$, effectively H plays the role of the Hubble constant. In this case one has the very simple form for the Ricci tensor: $R_b^a = -3H^2\delta_b^a$.

The solution of (2.7) was explicitly obtained in [18] where it was found that the de Sitter-like solution is possible, with the scale factor given by $a(t) = H^{-1}\cosh Ht$, or $a(t) = a_0 \exp Ht$, or $a(t) = H^{-1}\sinh Ht$, for closed, flat and open Universe respectively. So, we see that accelerating solution is possible in this theory, just as in the presence of the cosmological term. Moreover, it is clear that a wide class of models involving higher orders in curvatures will admit accelerated solutions as well. This result called interest to $f(R)$ gravity displaying it to be a possible candidate for a consistent explanation of cosmic acceleration. Afterwards, many cosmological solutions for various versions of the function $f(R)$ were obtained and observationally tested, some of these results will be discussed in the next section.

Now, let us discuss the problem of degrees of freedom in R^2-gravity. First of all, we note that there is a common difficulty characteristic for higher-derivative theories, either gravitational or not. Indeed, in any Lorentz-invariant theory with four derivatives, the propagator will be proportional to the momentum depending factor looking like:

$$f(k) = \frac{1}{k^2 - \frac{k^4}{M^2}}, \tag{2.8}$$

where M^2 is the energy scale at which the higher derivatives become important. It is clear that we can rewrite this factor as

$$f(k) = \frac{1}{k^2} - \frac{1}{k^2 - M^2}. \tag{2.9}$$

Therefore we see that this propagator actually describes two distinct degrees of freedom, the massive and the massless one. Moreover, these two contributions to the propagator have opposite signs (otherwise, if signs of these contributions are the same, the UV behavior of the propagator is not improved). Clearly it means that the Hamiltonian describing these two degrees of freedom is composed by two terms with opposite signs:

$$\mathcal{H} = \frac{1}{2}(\pi_1^2 + \partial_i\phi_1\partial_i\phi_1) - \frac{1}{2}(\pi_2^2 + \partial_i\phi_2\partial_i\phi_2 + M^2\phi_2^2). \tag{2.10}$$

We see that the energy is not bounded from below, hence, we cannot define a vacuum in the theory consistently, i.e. one can take energy from the system without any limitations, as from a well without a bottom. Moreover, actually it means that the spectrum of the theory describes free particles with negative energy which seems to be nonsense from the viewpoint of the common sense. Actually this is the simplest example of the so-called Ostrogradsky instability plaguing higher-derivative field theory models except of special cases, see a detailed discussion of this example and similar situations in [19]; a profound discussion of difficulties arising within the Hamiltonian formulation of these theories is given also in [20]. Moreover, in some cases the higher-derivative theories involve not only ghosts but even tachyons, for a specific sign of the higher-derivative term. Therefore the higher-derivative models including the R^2-gravity are treated as effective theories aimed for description of the low-energy dynamics of the theory (roughly speaking, for the square of momentum much less than the characteristic mass square M^2). However, it is necessary to note that higher-derivative terms naturally emerge as quantum corrections after the integration over some matter fields, see f.e. [21], so, the presence of higher-derivative terms within the effective dynamics in many field theory models including gravity is natural.

Within our R^2-gravity model, the presence of ghosts can be illustrated as follows. If one will extract only physical degrees of freedom, whose role is played by transverse-traceless parts of spatial components h_{ij}^{TT} of the metric fluctuation, expanded as $h_{ij}^{TT} = K_{ij} + F_{ij}$, and scalar fields, one will see that the quadratic action will look like [17]

$$\mathcal{L}_K = -\frac{\gamma}{4} K_{ij} \Box K_{ij} + \frac{\gamma}{4} F_{ij}(\Box + m_2^2) F_{ij} -$$
$$- \frac{1}{8} h^T [(8\beta - 3\alpha)\kappa^2 \Box + \gamma] \Box h^T + \dots, \qquad (2.11)$$

where $h_T = h_{ii} - \nabla^{-2} h_{ij,ij}$ is a trace part of h_{ij}. We see that here, K_{ij} and F_{ij} behave as two degrees of freedom, with one of them is massive and another is massless, and their signs are opposite. Hence, the ghost contributions emerge naturally. We see that the number of degrees of freedom is increased, besides of tensor modes we have also scalar ones, and each of them is contributed by usual and ghost ones (the contribution for the scalar h^T can be also split into usual and ghost parts).

Clearly, the natural question is—whether is it possible to deal with ghosts or even avoid their presence? There are several answers to this question. One approach is based on extracting so-called "benign" ghosts whose contribution can be controlled [22, 23]. Another approach is based on considering the theory where the propagator has a form of the primitive monomial rather than the product of monomials as in (2.8). The simplest manner to do it consists in treating of the Lagrangian involving only higher-derivative term with no usual two-derivative one. Within the gravity context it means that one introduces the so-called pure R^2 gravity (sometimes referred as "agravity" [24]) where the usual Einstein-Hilbert term is absent. This theory was introduced in [24] and got its further development in [25], with its action can be treated as the special limit of R^2 gravity: $S = \sqrt{|g|}(\beta R^2 + \kappa^{-2} R)$, with $\kappa^{-2} \to 0$. The propagator will be proportional to

$$G_{\mu\nu\rho\sigma}(k) = \frac{1}{6\beta} \frac{1}{k^4} P_{\mu\nu,\rho\sigma}^0, \qquad (2.12)$$

with $P_{\mu\nu,\rho\sigma}^0 = \frac{1}{3} P_{\mu\nu} P_{\rho\sigma}$, the $P_{\rho\sigma}$ is the usual transverse projector, and β is a coefficient at R^2. One can show that on the flat background, only scalar mode propagates [25]. It is clear that there is no ghosts in this theory (in [25] it is also argued with analysis of degrees of freedom). It is interesting to note that the Breit potential for this propagator displays confining behavior:

$$V(\vec{r}) = \int \frac{d^3 k}{(2\pi)^4} \frac{e^{i\vec{k}\cdot\vec{r}}}{\vec{k}^4} \propto |\vec{r}|. \qquad (2.13)$$

So, this theory has only one difficulty—it does not yield Einstein-Hilbert limit which was tested through many observations. Many aspects of the pure R^2-gravity are discussed in [26–29], see also references therein.

2.3 $f(R)$-Gravity

Clearly, the natural development of the idea of R^2 gravity will consist in the suggestion that the classical action can involve not only second but any degree (involving negative!) of the scalar curvature. Thus, the concept of $f(R)$ gravity was introduced. Its action is given by (2.1), with $f(R) = R + \gamma R^n$.

First of all, we can discuss the renormalizability of this theory along the same lines as in the previous section. It is easy to see that the term proportional to R^N (or, which is similar, to N-th degree or Riemann or Ricci tensors) is characterized by the degree of divergence ω, in the four-dimensional space-time given by

$$\omega = 4L - 2n(P - V) - 2N = (4 - 2n)L + 2n - 2N. \qquad (2.14)$$

Immediately we see that now discussion of the renormalizability is more involved than for $n = 2$ (the similar situation occurs for Horava-Lifshitz-like theories where increasing of the critical exponent z implies in growing not only of degree of momentum in the denominator of the propagator but also of numbers of derivatives in vertices). Actually, for any $n > 2$ one should classify possible divergences with various values of N for the given n. Many examples of quantum calculations in theories for various n, as well as in other higher-derivative gravity theories, including studies of one-loop divergences and running couplings are presented in [10], see also references therein. It is clear that the ghosts will arise for any polynomial form of $f(R)$ just as in the case of R^2-gravity, so, conceptually the quantum calculations for $n = 2$ and for $n > 2$ do not differ essentially (for discussion of renormalizability aspects of $f(R)$ gravity, see also [30]; also, in [31] the renormalizability of a higher-derivative gravity is argued with use of the BRST symmetry).

The main line of study of $f(R)$ gravity consists in a detailed investigation of its classical, especially cosmological aspects. The modified Einstein equations in this case look like

$$f'(R)R_{\mu\nu} - \frac{1}{2}g_{\mu\nu}f(R) + (g_{\mu\nu}\nabla_\lambda\nabla^\lambda - \nabla_\mu\nabla_\nu)f'(R) = 8\pi G T_{\mu\nu}. \qquad (2.15)$$

It is evident that the de Sitter/anti-de Sitter (dS/adS) space-times will be vacuum solutions of these equations yielding $f(R) = bR^2 + \Lambda$, with b being a constant. Then, to study the cosmological aspects, we can use the expressions for components of the Ricci tensor and the scalar curvature (1.22). In a whole analogy with (2.7) one can find that, if the $f(R)$ involves R^2 term, the corresponding cosmological equation, up to dimensionless factors, will be

$$\frac{\dot{a}^2 + k}{a^2} = \frac{1}{H^2}\left(\frac{\dot{a}^2 + k}{a^2}\right)^2 - \frac{1}{M^{2n}}\left(\frac{\dot{a}^{2n}}{a^{2n}} + \cdots\right), \qquad (2.16)$$

where H is the constant, accompanying the R^2 term, cf. (2.7), and M is the energy scale accompanying the higher curvature term. The dots in parentheses are for other terms with $2n$ time derivatives (if $k = 0$ they are all homogeneous, involving the same degrees of a in the numerator and in the denominator). It can be shown (see f.e. [20] and references therein), that in this theory, for any $n \geq 2$ the solutions are again presented by hyperbolic sine and cosine and exponential, just as in R^2 case [18]. We conclude that this theory describes well the inflationary epoch where the curvature of the Universe was large hence the higher-derivative contributions are important. In principle, in this earlier epoch one can use the action introduced in the manner of [25, 26] where the Einstein-Hilbert term is suppressed, and one chooses $f(R) = \gamma R^n$ as a reasonable approximation. At the same time, an interesting problem is—how one can adopt the form of the $f(R)$ to explain the actual accelerated expansion of the Universe, in the case where the curvature is very close to zero, so, R^n terms with $n > 1$ can be disregarded.

In [32], a bold departure from usual forms of the $f(R)$ function was proposed: this function was suggested to be

$$f(R) = R - \frac{\mu^4}{R}. \tag{2.17}$$

This model is sometimes referred as $1/R$ gravity. The quantum description of this theory near the flat background is problematic. However, it can be treated perturbatively in principle near some other background.

Let us discuss the equations of motion for this choice of $f(R)$. In the vacuum case ($T_{\mu\nu} = 0$), we have

$$\left(1 + \frac{\mu^4}{R^2}\right) R_{\mu\nu} - \frac{1}{2}\left(1 - \frac{\mu^4}{R^2}\right) R g_{\mu\nu} + (g_{\mu\nu}\Box - \nabla_\mu\nabla_\nu)\frac{\mu^4}{R^2} = 0. \tag{2.18}$$

For the constant scalar curvature, one finds

$$R_{\mu\nu} = \pm\frac{\sqrt{3}}{4}\mu^2 g_{\mu\nu}, \tag{2.19}$$

this is, (a)dS solution, and in the case of the negative sign, at $\mu \neq 0$ we indeed have an acceleration [20], so, this model allows to explain accelerated expansion for the constant curvature case.

Unfortunately, this model suffers from a tachyonic instability. Indeed, after taking the trace of (2.18) we find

$$-R + \frac{3\mu^4}{R} + 3\Box\left(\frac{\mu^4}{R^2}\right) = 0. \tag{2.20}$$

After we make a perturbation δR around an accelerated solution described by a constant negative curvature, i.e. $R = -\sqrt{3}\mu^2 + \delta R$, we find that the δR obeys the equation

$$- \delta R + \frac{2}{\sqrt{3}\mu^2}\Box\delta R = 0, \tag{2.21}$$

and in our signature $(+ - - -)$ this equation describes a tachyon. Actually, this instability is very weak since μ^2 is expected to be very small, for the consistency with observations, hence the first term in this equation is highly suppressed. It should be noted that for a non-zero density of the matter the instability is much worse, but adding the R^2 term into the action improves radically the situation [20]. Therefore this model was naturally treated as one of candidates for solving the dark energy problem. However, the model (2.17), in further works, was discussed mostly within the cosmological context (see also a discussion of asymptotic behavior of cosmological solutions in [32]).

Let us note some more issues related to $f(R)$ gravity. First, it was argued in [20] that the $f(R)$ gravity model is equivalent to a some scalar-tensor gravity. Indeed, let us for the first step define $f(R) = R + \bar{f}(R)$, so $\bar{f}(R)$ is a correcting term. Then, we introduce an auxiliary scalar field $\phi = 1 + \bar{f}'(R)$. Since this equation relates R and ϕ, it can be solved, so one obtains a dependence $R = R(\phi)$. As a next step, the potential looking like

$$U(\phi) = (\phi - 1)R(\phi) - \bar{f}(R(\phi)), \tag{2.22}$$

implying $U'(\phi) = R(\phi)$, is defined. As a result, the Lagrangian (2.1) turns out to be equivalent to

$$\mathcal{L}_E = \sqrt{|g|}(\phi R - U(\phi)). \tag{2.23}$$

Then, we carry out the conformal transformation of the metric:

$$\tilde{g}_{\alpha\beta} = \phi g_{\alpha\beta}, \quad \phi = \exp\left(\sqrt{\frac{4\pi G}{3}}\varphi\right), \tag{2.24}$$

therefore the Lagrangian is rewritten as

$$\mathcal{L}_E = \sqrt{|\tilde{g}|}\left(\frac{1}{16\pi G}\tilde{R} - \frac{1}{2}\tilde{g}^{ab}\partial_a\varphi\partial_b\varphi - V(\varphi)\right),$$
$$V(\varphi) = \frac{1}{16\pi G}U\left(\exp\left(\sqrt{\frac{4\pi G}{3}}\varphi\right)\right)\exp\left(-\sqrt{\frac{16\pi G}{3}}\varphi\right). \tag{2.25}$$

Therefore, the $f(R)$ gravity turns out to be equivalent to the general relativity with the extra scalar, i.e. to the scalar-tensor gravity. The form of the potential is therefore related with the form of the function $f(R)$.

Clearly, the natural question is about the possibility to obtain other important gravitational solutions within the $f(R)$ gravity context. First, for the Gödel metric (1.8), as well as for its straightforward generalization, that is, the Gödel-type metric (1.10), the scalar curvature is constant, hence the equations (2.15) are simplified drastically since the term involving covariant derivatives of $f(R)$ goes away, and the l.h.s. of these equations turns out to be a mere combination of constants. As we already noted, it was shown in [33] that both noncausal solutions, with $0 < m^2 < 4\Omega^2$, and causal ones, f.e. those ones with $m^2 \geq 4\Omega^2$ (i.e. the RT metric with $m^2 = 4\Omega^2$ is causal), are possible in this theory, with $f(R)$ is an arbitrary function of the scalar curvature, while to achieve causality, it is not sufficient to have only a relativistic fluid as in [4], and one must add as well a scalar matter—one should remind that since the Einstein equations are nonlinear, the solution generated by a sum of two sources is not equal to the sum of solutions generated by each source. As for the black holes, we strongly recommend the excellent book [34] where Schwarzschild-type BH solutions in $f(R)$ gravity are considered, see also [35] and references therein.

In [34], a wide spectrum of possible generalizations of $f(R)$ gravity was discussed, such as $f(R, \mathcal{L}_m)$ and $f(R, T)$ models, where $\mathcal{L}_m$ is the matter Lagrangian, and T is the trace of the energy-momentum tensor. However, within our study we will pursue another aim—we will suggest that the matter is coupled to the gravity in the usual form while the free gravity action depends on other scalars constructed on the base of the Riemann tensor and metric. This will be the subject of the next section.

2.4 Functions of Other Curvature Invariants

Let us suggest that instead of the function of the scalar curvature only, we have also functions of other scalars. There are many examples of studies of such models, so we discuss only some most interesting ones, the $f(R, Q)$ and $f(R, Q, P)$ gravities, the Lovelock gravity and the Gauss-Bonnet gravity. We note that, unlike the $f(R)$ gravity, these theories in general cannot be reduced to the scalar-tensor form where the Lagrangian is a sum of a term linear in the scalar curvature and a curvature-independent term depending on some extra scalar field.

We start our discussion from the $f(R, Q)$ gravity. In this theory, the Lagrangian is a function not only of the scalar curvature, but also of $Q = R_{\mu\nu}R^{\mu\nu}$, so,

$$S = \int d^4x \sqrt{|g|} f(R, Q) + S_m. \tag{2.26}$$

The equations of motion are found to look like [36, 37]

$$f_R R_{\mu\nu} - \frac{f}{2} g_{\mu\nu} + 2 f_Q R^{\beta}_{(\mu} R_{\nu)\beta} + g_{\mu\nu} \Box f_R - \nabla_{(\mu} \nabla_{\nu)} f_R +$$
$$+ \Box \left(f_Q R_{\mu\nu} \right) - 2 \nabla_\lambda \left[\nabla_{(\mu} \left(f_Q R^{\lambda}_{\nu)} \right) \right] + g_{\mu\nu} \nabla_\alpha \nabla_\sigma \left(f_Q R^{\alpha\sigma} \right) = \kappa^2 T^m_{\mu\nu}, \tag{2.27}$$

where $f_Q = \frac{\partial f}{\partial Q}$, $f_R = \frac{\partial f}{\partial R}$, and $T_{\mu\nu}^m$ is the energy-momentum tensor of the matter.

As an example, we consider the Gödel-type metric (1.10). One can show, that, unlike general relativity, such solutions are possible not only for dust but also for the vacuum (with non-zero cosmological constant), in particular, completely causal vacuum solutions are present [37]. Clearly, the solutions of this form are possible also for the presence of the matter given by the relativistic fluid and a scalar field. Again, as in [33], all Einstein equations will take the form of purely algebraic relations between density, pressure, field amplitude and constants from the gravity Lagrangian. As for the cosmological metric, the possibility of accelerating solutions can be shown just in the same manner as in the previous sections. Among other possible solutions in $f(R, Q)$ gravity, it is worth to mention Reissner-Nordström black holes [38] and wormholes [39]. Further generalization of this theory would consist in consideration of function not only of R and Q, but also of $P = R_{\mu\nu\alpha\beta} R^{\mu\nu\alpha\beta}$, with study of the corresponding theory called $f(R, Q, P)$ gravity is in principle not more difficult, see f.e. [40, 41]. One of subclasses of such theories is represented by so-called $f(R, \mathcal{G})$ theories, where $\mathcal{G}$ is the Gauss-Bonnet invariant,

$$\mathcal{G} = R^2 - 4R_{\mu\nu}R^{\mu\nu} + 4R_{\mu\nu\alpha\beta}R^{\mu\nu\alpha\beta} \equiv R^2 - 4Q + P, \qquad (2.28)$$

whose contribution to the classical action yields a total derivative in the four-dimensional space-time. The $f(R, \mathcal{G})$ theories have been treated mostly within the cosmological context (see f.e. [42]).

Returning to more generic $f(R, Q, P)$ theories, we note that in the paper [41], the consistency of the Gödel-type metrics in this class of theories has been studied for various examples of the function $f(R, Q, P)$. For example, one of interesting cases is $f(R, Q, P) = R + g(aR^2 + bQ + cP)$, with $g(z)$ is a some function of z, f.e. $g(z) = -\frac{\mu^{4n+2}}{z^n}$. In this situation, the RT metric [5], that is, the Gödel-type metric characterized by the relation $m^2 = 4\Omega^2$, was shown to be consistent with the modified Einstein equations, in the vacuum case, for certain relation between the vorticity Ω and the theory parameters a, b, c. The cases with nontrivial matter sources also can yield completely causal solutions, for certain relations between the parameters of the theory.

One more interesting example of modification of the purely geometric sector is presented by the so-called Ricci-inverse gravity [43]. Its action looks like

$$S = \frac{1}{\kappa^2} \int d^4x \sqrt{|g|} f(R, A), \qquad (2.29)$$

where $A = A_\mu^\mu$, with $A_{\mu\nu} = (R_{\mu\nu})^{-1}$ is the inverse Ricci tensor. This model can be treated as a further development of the $1/R$ gravity [32] discussed above. The $f(R, A)$ is an arbitrary function of R and A, in the simplest case, $f(R, A) = R - \alpha A$. Within [43], the Ricci-inverse gravity has been considered within the cosmological context, and the possibility of accelerated solution was proved (the simplest example noted in [43] is: $f(R, A) = R - \alpha A$, so, the Hubble parameter $H = (\alpha/3)^{1/4}$, that is, the de Sitter solution with $R = 12H^2$). Certainly, other, non-cosmological solutions in this model can be studied as well.

Now, let us make the next step—suggest that the dimension of the space-time is not restricted to be four but can be arbitrary. This step allows us to introduce the Lovelock gravity. Its key idea is as follows.

Let us consider the gravity model defined in the space-time of an arbitrary dimension [44], called the Lovelock gravity:

$$S = \int d^D x \sqrt{|g|}(c_0 \Lambda + c_1 R + c_2 \mathcal{G} + \cdots). \tag{2.30}$$

Here $c_0, c_1, c_2, \ldots$ are some constants possessing nontrivial dimensions. It is natural to suggest that they, up to some dimensionless numbers, are given by various degrees of the gravitational constant. Each term with $2n$ derivatives is topological, i.e. it represents itself as a total derivative at $D = 2n$, and identical zero in minor dimensions. We note that there is no higher derivatives of the metric in the action. This action is characterized the following properties displayed by the Einstein-Hilbert action: (i) the tensor $A_{\alpha\beta}$, the l.h.s. of the corresponding equations of motion, is symmetric; (ii) the covariant divergence of $A_{\alpha\beta}$ vanishes; (iii) the $A_{\alpha\beta}$ is linear in second derivatives of the metric.

The general form of the term with $2n$ derivatives in the Lagrangian contributing to (2.30) can be presented as [45]:

$$\mathcal{L}_n = \frac{1}{2^n} \delta^{\alpha_1 \ldots \alpha_{2n}}_{\beta_1 \ldots \beta_{2n}} R^{\beta_1 \beta_2}{}_{\alpha_1 \alpha_2} \ldots R^{\beta_{2n-1} \beta_{2n}}{}_{\alpha_{2n-1} \alpha_{2n}}, \tag{2.31}$$

where the $2n$-order Kronecker-like delta symbol is

$$\delta^{\alpha_1 \ldots \alpha_{2n}}_{\beta_1 \ldots \beta_{2n}} = \begin{vmatrix} \delta^{\alpha_1}_{\beta_1} & \ldots & \delta^{\alpha_1}_{\beta_{2n}} \\ \ldots & \ldots & \ldots \\ \delta^{\alpha_{2n}}_{\beta_1} & \ldots & \delta^{\alpha_{2n}}_{\beta_{2n}} \end{vmatrix}. \tag{2.32}$$

It is easy to check that at $n = 1$, we have the scalar curvature, and at $n = 2$, the Gauss-Bonnet term. The term with $n = 0$ is naturally treated as the cosmological constant. As a result, we can write down the action;

$$S = \frac{1}{\kappa^2} \int d^D x \sqrt{|g|} \sum_{0 \le n < D/2} \alpha_n \lambda^{2(n-1)} \mathcal{L}_n. \tag{2.33}$$

Here, zero order is for Λ, first—for R, second—for $\mathcal{G}$. The α_n are some numbers, and λ is a length scale, f.e. Planck length, it is given by κ in $D = 4$ where the κ^{-1} has a dimension of inverse length.

The l.h.s. of the modified Einstein equations looks like [45]

$$G_{\alpha\beta} = \sum_{0 \leq n \leq D/2} \alpha_n \lambda^{2(n-1)} G_{(n)\alpha\beta};$$

$$G_{(n)}{}^{\alpha}{}_{\beta} = -\frac{1}{2^{n+1}} \delta^{\alpha\alpha_1...\alpha_{2n}}_{\beta\beta_1...\beta_{2n}} R^{\beta_1\beta_2}{}_{\alpha_1\alpha_2} \cdots R^{\beta_{2n-1}\beta_{2n}}{}_{\alpha_{2n-1}\alpha_{2n}}. \tag{2.34}$$

It is clear that $G_{(0)\alpha\beta} = -\frac{1}{2}g_{\alpha\beta}$, $G_{(1)\alpha\beta} = G^E_{\alpha\beta} \equiv R_{\alpha\beta} - \frac{1}{2}Rg_{\alpha\beta}$ is the usual Einstein tensor. The r.h.s. of the modified Einstein equations is not modified within this approach, so we have $G_{\alpha\beta} = \kappa^2 T_{\alpha\beta}$.

It turns out to be that although the l.h.s. (2.34) of the modified Einstein equations is very complicated, these equations admit some exact solutions for an arbitrary space-time dimension, i.e. for the presence of terms with very high orders in curvatures. The most interesting cases are the maximally symmetric (anti) de Sitter space and the FRW cosmological metric.

In the (a)dS space, the Riemann curvature tensor is given by

$$R_{\alpha\beta\gamma\delta} = \frac{\sigma}{\lambda^2}(g_{\alpha\gamma}g_{\beta\delta} - g_{\alpha\delta}g_{\beta\gamma}), \tag{2.35}$$

with σ is a some number. In this case the vacuum equation yields $\sum_{0 \leq n < D/2} \beta_n \sigma^n = 0$, with $\beta_n = \frac{(D-1)!}{(D-2n-1)!}\alpha_n$, and this equation possesses some roots for σ (in general complex ones). Each value of σ allows to find the corresponding scalar curvature.

We can solve the modified Einstein equations also for the FRW metric (1.6):

$$R_{i0j0} = -g_{ij}\frac{\ddot{a}}{a}; \quad R_{ijkl} = \frac{\dot{a}^2 + k}{a^2}(g_{ik}g_{jl} - g_{il}g_{jk}), \tag{2.36}$$

defining $A = \frac{\lambda^2(\dot{a}^2+k)}{a^2}$, where as usual $k = -1, 0, 1$, the indices i, j, k, l take values $1, 2, 3$, so that the l.h.s. of modified Einstein equations yields

$$\lambda^2 G_{00} = \frac{1}{2} \sum_{0 \leq n < D/2} \beta_n A^n, \tag{2.37}$$

$$\lambda^2 G_{ij} = -\frac{1}{2} \frac{g_{ij}}{2(D-1)} \sum_{0 \leq n < D/2} \beta_n A^{n-1} \left[2n\lambda^2 \frac{\ddot{a}}{a} + (D - 2n - 1)A \right].$$

For the vacuum ($k = 0$) one immediately finds $A = const$ which implies exponential expansion. For the fluid, also there are hyperbolic and trigonometric solutions. So, we conclude that the Lovelock gravity is consistent with accelerating expansion of the Universe.

Concerning the general Lovelock theory, it must be noted that already third-order contributions to the action, those ones with six derivatives, imply very complicated equations of motion. The explicit expressions for initial terms of Lovelock Lagrangians up to fifteenth order (for which, the whole expression involves tens of millions of terms) can be found in [46].

Now, let us discuss the Gauss-Bonnet gravity in the arbitrary spacetime dimension, that is, the theory with the action

$$S = \int d^D x \sqrt{|g|} \left(\frac{1}{2\kappa^2} R + f(\mathcal{G}) + \mathcal{L}_m \right). \tag{2.38}$$

The equations of motion of this theory are

$$\frac{1}{\kappa^2} G_{\mu\nu} = 2T_{\mu\nu} + \frac{1}{2} g_{\mu\nu} f(\mathcal{G}) - 2F(\mathcal{G}) R R_{\mu\nu} + 4F(\mathcal{G}) R_\mu^\lambda R_{\nu\lambda} -$$
$$- 2F(\mathcal{G}) R_{\mu\lambda\rho\sigma} R_\nu^{\lambda\rho\sigma} - 4F(\mathcal{G}) R_{\mu\rho\sigma\nu} R^{\rho\sigma} + 2R \nabla_\mu \nabla_\nu F(\mathcal{G}) -$$
$$- 2R g_{\mu\nu} \nabla^2 F(\mathcal{G}) - 4R_\mu^\rho \nabla_\nu \nabla_\rho F(\mathcal{G})$$
$$- 4R_\nu^\rho \nabla_\mu \nabla_\rho F(\mathcal{G}) + 4R_{\mu\nu} \nabla^2 F(\mathcal{G}) + 4g_{\mu\nu} R^{\lambda\rho} \nabla_\lambda \nabla_\rho F(\mathcal{G}) -$$
$$- 4R_{\mu\nu\lambda\rho} \nabla^\lambda \nabla^\rho F(\mathcal{G})$$
$$\equiv 2T_{\mu\nu} + H_{\mu\nu}, \tag{2.39}$$

with $F(\mathcal{G}) = f'(\mathcal{G})$. As a simple example, we discuss the solution of this equation in the braneworld case, i.e. we consider the five-dimensional metric

$$ds^2 = g_{\mu\nu} dx^\mu dx^\nu = e^{2A(y)} \eta_{ab} dx^a dx^b - dy^2, \tag{2.40}$$

where we suggest that the indices μ, ν vary from 0 to 4, while a, b—from 0 to 3, and y is the extra (fourth) spacial coordinate, and $A(y)$ is called the warp factor [47, 48]. In [49], these equations have been solved for the case when the matter is given by the scalar field ϕ, so, $T_{ab} = \eta_{ab} e^{2A} (\frac{1}{2} \phi'^2 + V(\phi))$, and $T_{44} = \frac{1}{2} \phi'^2 - V(\phi)$, for the simplified situation $\mathcal{G} = const = \pm 120b^2$. Explicitly, for positive $B(= \mathcal{G}/120) = b^2$, the solution is

$$y + C = \frac{4}{5} \int \frac{dA'(A')^2}{b^2 - (A')^4}, \tag{2.41}$$

and for the negative $B = -b^2$—in the form

$$y + C = -\frac{4}{5} \int \frac{dA'(A')^2}{b^2 + (A')^4}. \tag{2.42}$$

In principle, there are more situations when the modified Einstein equations in the Gauss-Bonnet gravity can be solved. We note that the braneworld solutions could be found not only for Gauss-Bonnet gravity but for other gravity models including the already discussed $f(R)$ gravity, where both constant and nonconstant scalar curvature solutions were obtained (see f.e. [50, 51] and references therein), however, we do not discuss the details of braneworld solutions here because of the restricted volume of this review.

2.5   Comments on Massive Gravity

In a full analogy with the vector field theory, where the Proca Lagrangian is introduced as a generalization of the Maxwell one, one more natural extension of the Einstein gravity consists in constructing its massive analogue.[1] For the first step, we consider the linearized case described by the Fierz-Pauli Lagrangian (1.13), and add to it the simplest massive term [52]:

$$S_m = \frac{1}{2} \int d^4x (h_{\mu\nu} h^{\mu\nu} - h^2), \qquad (2.43)$$

and the relative coefficient (-1) between two terms is chosen in order to avoid arising of additional modes, in particular, ghost ones.

Some discussions of this term are presented in [52]. Among various conclusions, it is argued there that, first, the complete linearized massive gravity action given by the sum of (1.13) and (2.43) describes the massive field with spin 2, with no dynamical spin-0 modes are present. The theory includes 5 °C of freedom. Indeed, the l.h.s. of the equations of motion for this theory is given by the sum of the l.h.s. of the usual linearized Einstein equation (1.14) and the term $m^2(h_{\mu\nu} - \eta_{\mu\nu}h)$ arising from our mass term. Taking the divergence of these equations, we find $\partial_\mu h^{\mu\nu} - \partial^\nu h = 0$, and after simple manipulations we arrive at the constraints $h = 0$ and $\partial_\mu h^{\mu\nu} = 0$, which allow to rewrite the Lagrange equation in the standard Proca-like form $(\Box + m^2)h_{\mu\nu} = 0$. So, we have five constraints, and hence there are five degrees of freedom. Certainly, their interpretation is an additional problem. Moreover, there is one more difficulty related with this theory—its propagator involves terms quadratically increasing with the momenta in the UV limit, i.e., it displays the confinement-like behaviour. From the formal viewpoint, the reason for it consists in the fact that, since the Lagrangian is not gauge invariant, the corresponding propagator, instead of the standard transversal projector $P_{\mu\nu} = \eta_{\mu\nu} - \frac{p_\mu p_\nu}{p^2}$ includes the operator $\tilde{P}_{\mu\nu} = \eta_{\mu\nu} - \frac{p_\mu p_\nu}{m^2}$ (see discussion in [52]). Besides of this worsening the UV behaviour, the presence of the non-transversal $\tilde{P}_{\mu\nu}$ naturally implies in a singularity at $m = 0$, so, the zero mass limit of our massive gravity is not well defined (moreover, if one compares the propagators of two theories, it turns out to be that the analogue of the last term in (1.18) for the massive gravity carries the factor $\frac{1}{D-1}$ instead of $\frac{1}{D-2}$ in (1.18), which complicates the problem of the $m = 0$ limit of our theory even more). The fact that the massless limit of our theory is ill-defined is referred as van Dam-Veltman-Zakharov (vDVZ) discontinuity originally discussed in [53, 54]. This discontinuity has been proved to generate problems also within the parametrized post-Newtonian (PPN) expansion [52] (we note that the weak field gravitational potential in the theory displays the standard form $\propto \frac{e^{-mr}}{r}$).

---

[1] Here we discuss massive terms in their own sense, i.e. those ones which break the gauge symmetry, generating analogues of the Proca theory for gravity, and do not consider additive higher-derivative gauge covariant terms like R^2, $R_{\mu\nu}R^{\mu\nu}$, whose presence, as we noted in the Sect. 2.2 generate massive modes.

One more difficulty with the massless limit of the massive gravity has been noted in [52], namely, if we try to obtain Stückelberg-like extension of our mass terms in order to re-establish the gauge symmetry, so that, first, we introduce the extra vector field A_μ replacing $h_{\mu\nu} \to h_{\mu\nu} + \partial_\mu A_\nu + \partial_\nu A_\mu$ so that the term (2.43) becomes gauge invariant under transformations $\delta h_{\mu\nu} = \partial_\mu \xi_\nu + \partial_\nu \xi_\mu$, $\delta A_\mu = -\xi_\mu$, as a result, the action for the A_μ field of the form

$$S_A = \int d^4x \left(-\frac{m^2}{2} F_{\mu\nu} F^{\mu\nu} - 2m^2 (h_{\mu\nu} \partial^\mu A^\nu - h\partial_\mu A^\mu) \right), \qquad (2.44)$$

and now, in order to convert A_μ to a gauge field, we perform one more Stückelberg-like extension $A_\mu \to A_\mu + \partial_\mu \phi$, as a result, we arrive at the graviton-scalar mixing terms $-2m^2 (h_{\mu\nu} \partial^\mu \partial^\nu \phi - h\Box\phi)$ (we note that these terms reproduce the linearized limit of some possible galileons-gravity couplings, with ϕ playing the role of the galileon, see the discussion in the Sect. 3.4). As it is claimed in [52], this form of the mass terms for gravity can be extended to a full-fledged gravity through adding terms like $\frac{1}{D} R(h_{\mu\nu} h^{\mu\nu} - h^2)$, for a D-dimensional space-time, with the contractions are performed with use of some nontrivial background metric $g^{\mu\nu}$ (we note that presence of the second metric is rather characteristic for massive gravity models, see also f.e. [56]). One can show that the static spherically symmetric metric (3.20) solves the equations of motion, with the components of the metric tensor are proportional to $\frac{e^{-mr}}{r} P(\frac{1}{r})$, where $P(\frac{1}{r})$ is a polynomial function in $1/r$.

From another side, in a nonlinear massive gravity there are not five but six degrees of freedom, with the additional one is a ghost usually referred as the Boulware-Deser (BD) ghost [55]. To circumvent this problem, in [56] a bimetric theory, involving, besides of the physical metric $g_{\mu\nu}$, the additional non-dynamical one $f_{\mu\nu}$ was proposed (see [57] for a further development of this idea). The resulting action represents itself as a sum of the usual Einstein-Hilbert action and a some function of $\sqrt{g^{\mu\lambda} f_{\lambda\nu}}$. It was shown in [56] that in such theories, the arising of the extra mode is avoided.

There are other approaches to the massive gravity as well, some of them are presented in [58, 59]. Some historical review on the problem of graviton mass, together with related phenomenological estimations, also can be found in [59]. Here, we discuss in more details one of the possible scenarios where massive gravity can be introduced, that is, the softly higher-derivative massive gravity [60].

In that paper, the action (2.3), involving terms of the second order in curvature, is extended by adding the following non-local massive term:

$$S_m = -\frac{1}{2} \int d^D x \left(h_{\mu\nu} m^2(\Box) h^{\mu\nu} - h m^2(\Box) h \right), \qquad (2.45)$$

where as usual, $h = h^\mu_\mu$, and $m^2(\Box) = L^{2\alpha-2}(-\Box)^\alpha$ is the mass function, with the choice $\alpha < 1$, enforcing the mass term to dominate in the infrared limit (or, as is the same, at large distances), and L is the length scale. Actually, this term is naturally applied within the Dvali-Gabadadze-Porrati (DGP) braneworld scenario [61]. We note that this term can be introduced in an arbitrary space-time dimension. One can show that the spin-2 modes are characterized by the propagator proportional to $[\frac{\gamma}{2}\Box^2 + \Box - m^2(\Box)]^{-1}$ which cannot be presented as a product of two primitive multipliers. Thus, in general, ghosts cannot arise within this theory, except in certain cases, that is, $\alpha = 0$ and $\alpha = 1$.

Also, in this paper it is shown that the Newtonian gravitational potential (actually, the analogue of the Breit potential for scalar particles exchanging by gravitons), for the three-dimensional space-time and at $\gamma = 0$ (absence of higher derivatives) and $\alpha = 1/2$, at small distances displays logarithmic dependence on r, but at large distances reproduces the standard four-dimensional Newtonian asymptotics $\propto 1/r$, while for the four-dimensional space-time it decays at small distances as $1/r$, and at large distances as $1/r^2$, i.e. the DGP potential represents itself as a kind of interpolation between D-dimensional and $(D + 1)$-dimensional gravities (cf. [62]). The similar situation takes place for the four-dimensional space-time. If we consider the Lagrangian involving, besides the terms of second order in curvature, also topological Chern-Simons and Ricci-Cotton terms which carry multipliers with nontrivial mass dimensions, extra poles, and hence ghosts and tachyons, are possible [60].

It is natural to expect that other massive gravity models can be constructed as well.

2.6 Conclusions

We discussed various extensions of Einstein gravity characterized by modifications in the purely geometric sector. These modifications are based on adding new scalars representing themselves not only as various degrees of the scalar curvature but also as functions of higher order curvature invariants. We explicitly demonstrated that the R^2 gravity is all-loop renormalizable, and that the most important solutions of general relativity, such as cosmological FRW metric and Gödel metric continue to be solutions within modified gravity. Moreover, we showed that modifications of the purely geometric sector allow for accelerated cosmological expansion being thus examples of reasonable solutions for the dark energy problem, so that the problem of choosing a better modification of the gravity apparently can be solved in principle while the problem of choice for the most adequate modification of the gravity is actually more observational and experimental than theoretical.

Within this section we presented several other interesting results. First, we described the argumentation allowing for establishing the equivalence between modifications in the purely geometric sector and adjusting the action of the extra scalar field coupled to gravity, which implies that the $f(R)$ gravity is equivalent to the scalar-tensor gravity with an appropriate potential. Second, we discussed the $1/R$ terms whose form seems to be highly controversial since the observed curvature of the space-time is very small hence these terms are very large. Third, we considered gravity models depending on other curvature invariants. Fourth, we considered possible generalizations of the gravity consistent within the extra dimensions concept. Also, we presented a brief review of various massive extensions for gravity. We note that many of them use an auxiliary metric thus representing themselves as some kinds of bimetric gravity models. Besides this, it is worth mentioning that there is an essentially three-dimensional massive gravity model—so-called new massive gravity, also called Bergshoeff-Hohm-Townsend (BHT) gravity [63]. Its action looks like

$$S = \frac{1}{\kappa^2} \int d^3x \sqrt{|g|} \left(R + \frac{1}{m^2} \left(R_{\mu\nu} R^{\mu\nu} - \frac{3}{8} R^2 \right) \right), \tag{2.46}$$

with the linearized equations of motion are

$$(\Box + m^2) G_{\mu\nu}^{lin} = 0, \tag{2.47}$$

with $G_{\mu\nu}^{lin}$ is the linearized Einstein tensor given by the l.h.s. of (1.14), so, we have the Klein-Gordon-like equation perfectly corresponding to massive fields. Various aspects of this theory were studied in [64], for example, consistency of (a)dS-like, static axially symmetric, and some black hole solutions. Certainly, more effects in various massive gravity theories still must be studied.

Further development of a general gravity model consists in the idea that for the complete description of the gravity it is not sufficient to study only the metric, so that the gravity model should be extended through adding some other fields which are necessary being fundamental ingredients of the complete theory. So, one must consider scalar-tensor and vector-tensor gravity models. We will consider some examples of these models in our next chapters.

Chapter 3
Scalar-Tensor Gravities

3.1 General Review

In the previous chapter, we demonstrated that the modifications of the purely geometric sector allow for obtaining interesting results, in particular, for a consistent
explanation of the cosmic acceleration. At the same time, we noted that $f(R)$ gravities are dynamically equivalent to some gravity models whose action is given by the
sum of the usual Einstein term and a new term depending on the extra scalar field
[20]. This field, being related with the function of curvature, evidently cannot be
associated with the matter; hence, it is natural to suggest that the complete description of gravity is given by the composition of the dynamical metric tensor and this
scalar field, so we have the scalar-tensor gravity model. Another motivation for a
scalar-tensor gravity arises from quintessence models in cosmology which involve a
very light scalar field called the quintessence field and are known to explain accelerated expansion of the Universe as well as the cosmological constant which therefore
implied active application of the quintessence field within the inflationary context
[65]. The advantage of the quintessence in comparison with the cosmological constant consists in the fact that the very tiny mass of the quintessence field (estimated
to be about 10^{-33} eV [66]) is much more reasonable from the theoretical viewpoint
than the presence of an extremely small cosmological constant implying the famous
cosmological constant problem, since even the massless scalar fields are physically
consistent.

While the quintessence is well discussed now (see, for example, [66] and references therein), there are other interesting manners to introduce new scalar fields in
the gravity; moreover, while the quintessence field is treated as a matter, the scalar
fields introduced within these approaches are interpreted as ingredients of the complete description of the gravity rather than the matter. One of these manners is the
Brans-Dicke gravity where the gravitational constant whose negative dimension is
responsible for the non-renormalizability of the gravity is suggested to be not a fundamental constant but a function of a slowly varying fundamental scalar field. Another
one is the four-dimensional Chern-Simons modified gravity where the pseudoscalar

A. Petrov et al., *Introduction to Modified Gravity*,
SpringerBriefs in Physics, https://doi.org/10.1007/978-3-031-46634-2_3

field allows to implement the CPT (and in certain cases Lorentz) symmetry breaking in the gravity context. And actually, one more model is intensively discussed in this context, that is, the Galileons model. Namely, these theories will be considered in this chapter.

3.2 Chern-Simons Modified Gravity

3.2.1 The Four-Dimensional Chern-Simons Modified Gravity Action

The three-dimensional Chern-Simons (CS) term has been originally introduced in the paper [67] within the context of electrodynamics, as an example of a term conciliating gauge invariance with a non-zero mass. It has been immediately generalized to the non-Abelian case, so, the CS Lagrangian looks like

$$\mathcal{L}_{CS}^A = \mathrm{tr}\,\epsilon^{\mu\nu\lambda}(A_\mu \partial_\nu A_\lambda + \frac{2}{3} A_\mu A_\nu A_\lambda), \tag{3.1}$$

where $A_\mu = A_\mu^a T^a$ is the Lie-algebra valued gauge field, T^a are the generators of the Lie group, and f^{abc} are the structure constants. In the gravity case, the role of the gauge field is played by the connection, and the three-dimensional gravitational CS term reads as [67, 68]

$$S_{CS} = \frac{1}{2\kappa^2\mu} \int d^3x \epsilon^{\mu\nu\lambda}(\Gamma_{\mu a}^{\ b}\partial_\nu\Gamma_{\lambda b}^{\ a} + \frac{2}{3}\Gamma_{\mu a}^{\ b}\Gamma_{\nu b}^{\ c}\Gamma_{\lambda c}^{\ a}). \tag{3.2}$$

In metric-affine theories of gravity, the connection and metric are assumed to be independent entities, however, this general scenario is just considered in Sect. 7. We here assume the connection to be metric-compatible. Here, the $\epsilon^{\mu\nu\lambda}$, which can take values $1, 0, -1$, is the usual Levi-Civita symbol, not the covariant one. Varying the CS term with respect to the metric, one finds

$$\delta S_{CS} = -\frac{1}{\kappa^2\mu} \int d^3x C^{\mu\nu}\delta g_{\mu\nu}, \tag{3.3}$$

where

$$C^{\mu\nu} = -\frac{1}{2\sqrt{|g|}}\epsilon^{\mu\alpha\beta}\nabla_\alpha R_\beta^\nu + (\mu \leftrightarrow \nu) \tag{3.4}$$

is the three-dimensional Cotton tensor. It is evidently symmetric and traceless. The μ is a constant of the mass dimension 1. So, the modified Einstein equations look like

$$G^{\mu\nu} + \frac{1}{\mu}C^{\mu\nu} = 0. \tag{3.5}$$

It is useful also to write the linearized form of the gravitational Chern-Simons action obtained from (3.2) under the replacement $g_{\mu\nu} = \eta_{\mu\nu} + \kappa h_{\mu\nu}$:

$$S^{(0)} = -\frac{1}{2\mu} \int d^3x\, h^{\mu\nu}\epsilon_{\alpha\mu\rho}\partial^\rho(\Box\eta_{\gamma\nu} - \partial_\gamma\partial_\nu)h^{\gamma\alpha}. \tag{3.6}$$

We see that this action is, first, explicitly gauge invariant under usual linearized gauge transformations $\delta h_{\mu\nu} = \partial_\mu\xi_\nu + \partial_\nu\xi_\mu$; second, it involves higher derivatives. However, after obtaining the equations of motion for the full linearized action formed by the sum of the terms (1.13) and (3.6), one finds that the physical degrees of freedom satisfy the second-order equation [67], with their propagator behaving as $(\Box + \mu^2)^{-1}$; thus, in the three-dimensional CS modified gravity, there is no problems with negative-energy states discussed in the previous chapter. A similar situation occurs in the four-dimensional case as well.

The generalization of this theory to the four-dimensional case turns out to be straightforward; however, in this case, similar to the electrodynamics, this generalization essentially involves the CPT (and in certain cases Lorentz) symmetry breaking. From a formal viewpoint, such a generalization for the linearized theory is performed through replacement of $\epsilon^{\mu\nu\lambda} \to b_\rho\epsilon^{\rho\mu\nu\lambda}$, with b_ρ being a constant vector, which allows to promote the CS term to the Lorentz-violating (LV) Carroll-Field-Jackiw (CFJ) term which in the Abelian case looks like [69]

$$\mathcal{L}_{CFJ} = \epsilon^{\rho\mu\nu\lambda}b_\rho A_\mu\partial_\nu A_\lambda. \tag{3.7}$$

In principle, such a replacement of the three-dimensional Levi-Civita symbol by the four-dimensional one contracted with a vector already allows to write down the four-dimensional gravitational CS term [68]:

$$\mathcal{L}_{CS,grav} = \int d^4x\, \epsilon^{\rho\mu\nu\lambda}b_\rho(\Gamma_{\mu a}{}^{b}\partial_\nu\Gamma_{\lambda b}{}^{a} + \frac{2}{3}\Gamma_{\mu a}{}^{b}\Gamma_{\nu b}{}^{c}\Gamma_{\lambda c}{}^{a}), \tag{3.8}$$

with its linearized form being

$$S^{(0)} = -\frac{1}{2} \int d^4x\, h^{\mu\nu}\epsilon_{\alpha\mu\rho\lambda}b^\lambda\partial^\rho(\Box\eta_{\gamma\nu} - \partial_\gamma\partial_\nu)h^{\gamma\alpha}. \tag{3.9}$$

We note that this action is invariant under the same linearized gauge transformations $\delta h_{\mu\nu} = \partial_\mu\xi_\nu + \partial_\nu\xi_\mu$. Now, it is very interesting to discuss some motivations for this term.

First of all, already in 1984, much time before the interest to Lorentz-CPT breaking strongly increased, the gravitational anomalies were discussed in [70], where the topological current K^μ was introduced, with its explicit form being

$$K^\rho = 2\epsilon^{\rho\mu\nu\lambda}(\Gamma_{\mu a}^{\ b}\partial_\nu\Gamma_{\lambda b}^{\ a} + \frac{2}{3}\Gamma_{\mu a}^{\ b}\Gamma_{\nu b}^{\ c}\Gamma_{\lambda c}^{\ a}), \tag{3.10}$$

with its divergence being

$$\partial_\rho K^\rho = \frac{1}{2}\epsilon^{\mu\nu\alpha\beta}R_{\mu\nu\gamma\delta}R_{\alpha\beta}^{\ \ \gamma\delta} \equiv {}^*RR. \tag{3.11}$$

We note that the three-dimensional gravitational Chern-Simons term, up to overall multiplier, is equal to the K^3 component, i.e. the component of this current directed along "extra", z axis.

It is clear that the integral from (3.11) over the space-time is a surface term. To include it into the action in a consistent form, one should introduce a new field ϑ called the CS coefficient. As a result, one can add to the usual Einstein-Hilbert action the new term proportional to ϑ which will be called the CS action S_{CS} (cf. [68]):

$$S_{CS} = \frac{1}{2\kappa^2}\int d^4x(-\frac{1}{2}v_\mu K^\mu) = \frac{1}{2\kappa^2}I_{CS};$$
$$S_{EH+CS} = \frac{1}{2\kappa^2}\int d^4x(\sqrt{|g|}R + \frac{1}{4}\vartheta\,{}^*RR). \tag{3.12}$$

Here, $v_\mu = \partial_\mu\vartheta$ is a vector. We note that in principle this vector is rather a function of space-time coordinates than a constant; hence, in general the gravitational CS term breaks the CPT symmetry. However, the ϑ can be treated as an external, but not dynamical, field, therefore one can choose v_μ to be the constant vector. This immediately implies the Lorentz symmetry breaking, therefore in this case the four-dimensional CS modified gravity (CSMG), whose action is given by the second equation in (3.12), turns out to be one of the first examples of the gravity models incorporating Lorentz symmetry breaking.

The equations of motion for the CSMG can be easily obtained. Varying the CS term I_{CS}, defined by the first equation in (3.12), we get (see [68] for details)

$$\delta I_{CS} = \int d^4x\sqrt{|g|}C^{\mu\nu}\delta g_{\mu\nu}, \tag{3.13}$$

with $\varepsilon^{\alpha\beta\gamma\delta} = \frac{\epsilon^{\alpha\beta\gamma\delta}}{\sqrt{|g|}}$ being a Levi-Civita tensor (not a simple symbol!), and

$$C^{\mu\nu} = -\frac{1}{2}[v_\sigma(\varepsilon^{\sigma\mu\alpha\beta}\nabla_\alpha R_\beta^\nu + \varepsilon^{\sigma\nu\alpha\beta}\nabla_\alpha R_\beta^\mu) + v_{\sigma\tau}({}^*R^{\tau\mu\sigma\nu} + {}^*R^{\tau\nu\sigma\mu})] \tag{3.14}$$

being the Cotton tensor, and $v_{\sigma\tau} = \nabla_\sigma v_\tau$. One can check that the covariant divergence of the Cotton tensor is proportional to the invariant ${}^*RR$:

$$\nabla_\mu C^{\mu\nu} = \frac{1}{8}v^\nu\,{}^*RR. \tag{3.15}$$

This divergence plays a crucial role when the modified Einstein equations are considered. Their explicit form is

$$G^{\mu\nu} + C^{\mu\nu} = \kappa^2 T^{\mu\nu}, \tag{3.16}$$

so, due to Eq. (3.15), we find that the conservation of the energy-momentum tensor requires the vanishing of the divergence of the Cotton tensor, which, according to (3.15), yields an additional consistency condition called the Pontryagin constraint:

$$^*RR = 0, \tag{3.17}$$

which must be checked for any solution. However, since in many cases, including, among others, the rotational symmetry, the curvature tensor has the structure $R_{[ab][ab]}$, i.e. its only non-zero components are $R_{0101}, R_{0202}, \ldots$, this consistency condition will be automatically satisfied in these cases.

The further extension of the CSMG was carried out by assuming the nontrivial dynamics for the ϑ CS coefficient. The key idea is as follows [71]: we assume that the action of CSMG includes the kinetic term for ϑ, looking like

$$S = \frac{1}{2\kappa^2} \int d^4x \sqrt{|g|}(R + \frac{1}{2}\nabla^\mu\vartheta\nabla_\mu\vartheta - V(\vartheta) - \frac{1}{\alpha}\vartheta^*RR), \tag{3.18}$$

with now $^*RR \equiv \frac{1}{2}\varepsilon^{\mu\nu\alpha\beta}R_{\mu\nu\gamma\delta}R_{\alpha\beta}{}^{\gamma\delta}$, i.e. it is redefined with the Levi-Civita tensor $\varepsilon^{\mu\nu\lambda\rho} = \frac{\epsilon^{\mu\nu\lambda\rho}}{\sqrt{|g|}}$, and instead of the Pontryagin constraint (3.17), one has the equation of motion for ϑ:

$$^*RR = -\alpha(\Box\vartheta + \frac{\partial V}{\partial\vartheta}). \tag{3.19}$$

If we have a metric consistent within the non-dynamical CS framework with a specific ϑ, it is consistent in the dynamical case if the r.h.s. of this equation is zero. Then, the ϑ field generates the additional contribution to the energy-momentum tensor $T^{\mu\nu}$, and, hence, to the r.h.s. of (3.16).

Now, we present, first, some classical solutions for the CSMG, and, second, the methodology allowing the gravitational CS term as a quantum correction.

3.2.2 Classical Solutions

So, our task will consist in solving Eq. (3.16) with the additional condition (3.17). As a first example, we consider a static spherically symmetric metric [68]:

$$ds^2 = N^2(r)dt^2 - A^2(r)dr^2 - r^2(d\theta^2 + \sin^2\theta d\phi^2). \tag{3.20}$$

This is a very broad class of metrics including Schwarzschild, Reissner-Nordström and many other metrics. As we already said, in this case the non-zero components of the curvature tensor are $R_{[ab][ab]}$, so, the consistency condition (3.17) is automatically satisfied. For this metric, one has only non-zero components of the spatial Ricci tensor (i.e. the one corresponding to the spatial part of our metric): $R_r^r = \frac{A'}{rA^2}$, $R_\theta^\theta = \frac{1}{r^2}(1 - \frac{1}{A}) + \frac{A'}{rA^2}$. Then, we can consider the vacuum case $T^{\mu\nu} = 0$, and choose the vector $v_\mu = \partial_\mu\vartheta$ to be purely time-like, $v_\mu = (\frac{1}{\mu}, \vec{0})$, with $\mu = const$, i.e. $\vartheta = \frac{t}{\mu}$. In this case, the components C^{00} and $C^{0i} = C^{i0}$ of the Cotton tensor immediately vanish [68]. A bit more involved calculation (see details in [68]) allows to show that the C^{ij} components also vanish. As a result, we conclude that the spherically symmetric static solutions of the usual Einstein equations solve the modified equations (3.16) as well. It is clear that if one suggests the ϑ to be dynamical, Eq. (3.19) for ϑ will be satisfied if the potential is zero, and $\vartheta = \frac{t}{\mu}$. We note that this choice for ϑ is a particular case of the expression $\vartheta = k_\mu x^\mu$ used within studies of the Lorentz symmetry breaking in CSMG which we will discuss further.

Moreover, it has been shown in [71] that all, even non-static ones, spherically symmetric metrics given by

$$ds^2 = g_{\mu\nu}(x^\lambda)dx^\mu dx^\nu + \Phi^2(x^\rho)(d\theta^2 + \sin^2\theta d\phi^2), \qquad (3.21)$$

so that the coordinates on the sphere are x^i, and x^μ are two remaining coordinates (one of them is necessarily time-like), solve the modified Einstein equations (3.16) for

$$\vartheta = F(x^\mu) + \Phi(x^\mu)G(x^i), \qquad (3.22)$$

where $G(x^i)$ and $F(x^\gamma)$ are the arbitrary functions of sphere coordinates and remaining coordinates respectively, and Φ is defined in (3.21). The class of spherically symmetric metrics (3.21) involves not only the static ones (3.20) but also many other metrics, including the FRW cosmological metric (the cosmological aspects of CSMG were also discussed in many papers, for example, in [72]). Some types of metrics with cylindrical symmetry were also shown in [71] to be consistent within the CSMG.

Now, let us discuss the consistency of the Gödel-type metric (1.10) in CSMG. We consider the equations of motion (3.16) in the tetrad base, following [73, 74].

In the non-dynamical case [73], with appropriate choice of units, capital Latin indices are related to the tetrad ones, the equations (3.16) imply

$$R_{AB} + C_{AB} = \kappa^2(T_{AB} - \frac{1}{2}\eta_{AB}T) + \Lambda\eta_{AB}; \qquad (3.23)$$

$$C^{AB} = -\frac{1}{2}[\varepsilon^{CADE}(\nabla_D R_E^B)\partial_C\vartheta + {}^*R^{EAFB}\nabla_E\nabla_F\vartheta] + (A \leftrightarrow B).$$

Further, within this section, we choose $\kappa^2 = 1$. The divergence of modified Einstein equations is

$$\nabla_A C^{AB} = \frac{1}{8} {}^*RR\partial^B \vartheta. \tag{3.24}$$

In tetrad base, the components of Ricci tensor for Gödel-type metric are constant, which is an essential advantage of this base. Actually, one has

$$R_{00} = 2\Omega^2, \quad R_{11} = R_{22} = 2\Omega^2 - m^2, \quad R = 2(\Omega^2 - m^2). \tag{3.25}$$

Now let us choose the matter. We have its three most important examples [5, 73, 74]:

(i) Fluid, $T_{AB} = (\rho + p)u_A u_B + p\eta_{AB}$, $u^A = (1, 0, 0, 0)$, $T_{00} = \rho$ and $T_{11,22,33} = p$.
(ii) Scalar, $\psi = s(z - z_0)$, $T_{00,33} = \frac{s^2}{2}$ and $T_{11,22} = -\frac{s^2}{2}$.
(iii) Electromagnetism, $F_{03} = -F_{30} = e\sin[2\Omega(z - z_0)]$, $F_{12} = -F_{21} = -e\cos(2\Omega(z - z_0))$, $T_{00,11,22} = \frac{e^2}{2}$ and $T_{33} = -\frac{e^2}{2}$.

The matter can be presented by the composition of these three types. Then, the non-zero components of the Cotton tensor in this base look like

$$C_{00} = 2\frac{\partial\vartheta}{\partial z}\Omega(4\Omega^2 - m^2); \quad C_{11} = C_{22} = \frac{1}{2}C_{00};$$

$$C_{01} = -\frac{1}{2}\frac{\partial^2\vartheta}{\partial z\partial t}\frac{H}{D}(4\Omega^2 - m^2);$$

$$C_{02} = -\frac{1}{2}\frac{\partial^2\vartheta}{\partial z\partial r}(4\Omega^2 - m^2);$$

$$C_{03} = -\frac{1}{2}\frac{\partial\vartheta}{\partial t}\Omega(4\Omega^2 - m^2);$$

$$C_{13} = -\frac{1}{2}\frac{\partial^2\vartheta}{\partial t^2}\frac{H}{D}(4\Omega^2 - m^2);$$

$$C_{23} = \frac{1}{2}\frac{\partial^2\vartheta}{\partial r\partial t}(4\Omega^2 - m^2). \tag{3.26}$$

It is clear that the Cotton tensor is traceless, $C_A^A = 0$. To cancel the off-diagonal components of C_{AB}, we choose $\vartheta(z) = b(z - z_0)$ which matches the suggestion given above that the vector $v_M = \partial_M\vartheta$ is constant, which will be further used to study the Lorentz symmetry breaking. We introduce also $k = b\Omega$, and require $4\Omega^2 \neq m^2$, to consider solutions different from the RT metric [5] for which the Cotton tensor is evidently annihilated, so that the situation reduces to the usual GR case.

The system of the modified Einstein equations (for 00, 11=22, 33 components respectively) looks like:

$$2\,\Omega^2 + 2\,b\Omega(4\,\omega^2 - m^2) = \frac{1}{2}\,e^2 + \frac{1}{2}\,\rho - \Lambda + \frac{3}{2}\,p, \qquad (3.27)$$

$$2\,\Omega^2 - m^2 + b\Omega(4\,\Omega^2 - m^2) = \frac{1}{2}\,e^2 - \frac{1}{2}\,p + \Lambda + \frac{1}{2}\,\rho,$$

$$0 = -\frac{1}{2}\,e^2 - \frac{1}{2}\,p + s^2 + \Lambda + \frac{1}{2}\,\rho.$$

We note, that, just as in the GR case (cf. [5]), this system is a purely algebraic one. Let us solve these equations. After some manipulations, we arrive at equations for m^2 and Ω^2, with $k = b\Omega$ (we note that at $b = 0$, the usual GR solution is reproduced since in this case, $\vartheta = 0$!):

$$(2 + 8\,k)\Omega^2 - 2\,km^2 = \rho + s^2 + p, \qquad (3.28)$$

$$(2 + 4\,k)\,\Omega^2 - (1 + k)\,m^2 = -s^2 + e^2. \qquad (3.29)$$

One of the interesting new results having no GR analogue is the vacuum non-causal solution $m^2 = \Omega^2$, $b = -\frac{1}{3\Omega}$ and $\Lambda = 0$. Some other interesting conclusions of the above system are that, unlike the general relativity, the hyperbolic causal solutions are possible in CS modified gravity, and that trigonometric and linear solutions can arise only for a non-zero electromagnetic field [73].

If one suggests the CS coefficient to be dynamical, more new solutions having analogues neither in GR nor for the case of the non-dynamical CS coefficient are possible, see details in [74], with again the modified Einstein equations reducing to the algebraic equations involving some extra additive terms in comparison with (3.27). In particular, one can have a vacuum *causal* solution, where only the cosmological constant is non-zero, while density, pressure and all fields are zero.

At the same time, it is necessary to emphasize that not any solution consistent in the GR will be consistent also in CSMG. The paradigmatic example is the Kerr metric which fails to solve equations of motion in CSMG [68, 75]. It has been shown then in [76] that, to satisfy the modified Einstein equations in the dynamical CSMG, the Kerr metric should be also modified, by adding the ϑ-dependent terms, with the equations of motion being afterwards solved order by order in ϑ. Clearly, studies of consistency of various metrics possessing no rotational symmetry within the CSMG represent an open problem.

To close the discussion of the classical solutions, it is necessary to discuss the propagation of the plane waves. Similar to Sect. 2.2, we introduce the transverse-traceless components $h_{ij}^{T,T}$ which are the only physical variables in the theory (so, there are only two independent components, that is, if the plane wave propagates, for example, along x_3, we have only $h_{11} = -h_{22} = T$ and $h_{12} = h_{21} = S$).

In this case, for the time-like vector $v_\mu = (\mu^{-1}, 0, 0, 0)$, the quadratic Lagrangian takes the form:

$$L_2 = -\frac{1}{4}h_{ij}^{TT}\Box h_{ij}^{TT} + \frac{1}{4\mu}\epsilon^{ijk}h_{il}^{TT}\Box\partial_k h_j^{l,TT} + \dots, \qquad (3.30)$$

where dots are for physically irrelevant (non-propagating) degrees of freedom.

The corresponding linear equation of motion is

$$-\frac{1}{2}\Box h_{TT}^{ij} + \frac{1}{2\mu}\epsilon^{ilk}\Box\partial_k h_{l,TT}^{j} = 0. \tag{3.31}$$

As a result, one immediately concludes that the dispersion relation is the usual one, $k_0^2 = \vec{k}^2$, and both polarizations propagate with the speed of light.

The natural question is—what is the difference between these polarizations? A more careful analysis [68] shows that, for plane waves proportional to $e^{i\omega t - ikz}$, one finds that there are two basic (circular) polarizations $T = iS$ and $T = -iS$, with their intensities proportional to $(1 + \frac{k}{\mu})^{-2}$ and $(1 - \frac{k}{\mu})^{-2}$, respectively. This difference of intensities can be treated as a consequence of parity breaking.

It should be noted that if we consider, instead of the CS term, the one-derivative additive term $h_{\mu\nu}\epsilon^{\lambda\alpha\mu\rho}\theta_\lambda\partial_\rho h_\alpha^\nu$, with θ^λ being a space-like vector, we will have two polarizations with physically consistent dispersion relations $E = \pm\theta + \sqrt{p^2 + \theta^2}$, so, in this case the velocities differ from the speed of light [77]. However, this term is not gauge invariant, which, within the gravity context, means that it breaks the general covariance.

3.2.3 Perturbative Generation

The special interest is attracted to the gravitational CS term within the context of study of the Lorentz symmetry breaking. The main reason consists in the fact that, besides the CPT symmetry breaking, for a special choice of the CS coefficient $\vartheta = b_\mu x^\mu$, where b_μ is a constant vector (as we already noted in the previous subsection, this choice is consistent with the Gödel-type solutions), the CS term displays Lorentz symmetry breaking, taking the form (3.8), or, for the weak field, the linearized form (3.9). Therefore, the natural idea consists in a generation of this term as a perturbative correction, similar to the generation of the CFJ term in the extended QED; see, for example, [78]. This similarity is supported by a natural analogy between the gravitational anomalies [70] and the Adler-Bell-Jackiw (ABJ) anomaly [79]. Moreover, it follows from [80] that this anomaly is deeply related with the ambiguity of results; therefore, it is natural to expect the ambiguity of the gravitational CS term as well.

So, one can start with the action of spinors coupled to gravity, where the Lorentz-breaking vector b_μ is introduced:

$$S = \int d^4x\, e\bar{\psi}\,(i\slashed{\partial} - m - \slashed{b}\gamma_5 + \slashed{\omega})\psi, \tag{3.32}$$

where $\slashed{b} = b^\mu e_\mu^a \gamma_a$, and $\omega_\mu = \frac{1}{4}\omega_{\mu bc}\sigma^{bc}$ is a (Riemannian) connection. We note that the CS term dominates in the limit $m \to 0$ while the one-derivative term discussed in [77] vanishes in this limit. The corresponding one-loop effective action is given by the following trace of the logarithm:

$$\Gamma^{(1)} = i \operatorname{Tr} \ln(i\slashed{\partial} - m - \slashed{b}\gamma_5 + \phi). \tag{3.33}$$

Just the same approach was used in [78] for the Lorentz-breaking extension of QED. In the weak gravity case, we can use the approximation $e_{\mu a} \simeq \eta_{\mu a} + \frac{1}{2}h_{\mu a}$. This trace of the logarithm, however, can be calculated both in the weak field case and in the full-fledged gravity case, with use of the Feynman diagrams or of the proper-time method.

It is interesting that, similar to the CFJ term, the four-dimensional gravitational CS term is ambiguous, i.e. the results for it depend on the calculation scheme. So, within all these approaches, the linearized gravitational CS term

$$S_{CS} = C \int d^4 x\, h_{\mu\nu}\epsilon^{\mu\rho\kappa\lambda}b_\kappa \partial_\lambda \left(\Box h_\rho{}^\nu - \partial^\nu \partial^\sigma h_{\rho\sigma} \right), \tag{3.34}$$

or its full-fledged analogue (3.8) multiplied by $2C$, was shown to arise, with the constant C depending on the method of computation. So, in [81], where the calculations were carried out in the weak gravity case with the use of the Feynman diagrams constructed for the action (3.32), it was found that $C = \frac{1}{192\pi^2}$. Further, in [82], this scheme has been realized for the finite temperature case where the zero component of the internal momentum is supposed to be discrete, $k_0 = (2n + 1)\pi T$, so that the result is

$$S_{CS} = \int d^4 x\, h_{\mu\nu} \left[\frac{1}{192\pi^2} \epsilon^{\rho\mu\kappa\lambda} b_\kappa \partial_\lambda \left(\Box h_\rho{}^\nu - \partial^\nu \partial^\sigma h_{\rho\sigma} \right) \right. \tag{3.35}$$
$$\left. + \frac{T^2}{12} b_0 \epsilon^{\rho\mu\kappa\lambda} u_\kappa \partial_\lambda \left(\frac{\partial_0 \partial^\nu}{\Box} - u^\nu \right) \left(\frac{\partial_0 \partial^\sigma}{\Box} - u^\sigma \right) h_{\rho\sigma} \right],$$

i.e. it looks like a sum of the zero-temperature result (3.34) and the additive term proportional to T^2.

In [83], where the proper-time method has been used for the full-fledged gravity, the result was found in the form (3.8), with $C = \frac{1}{128\pi^2}$. Finally, in [84] it has been argued that due to the arbitrariness in defining conserved currents within the functional integral approach, the constant C is actually completely ambiguous. A similar situation occurs in QED [85]. However, the ambiguity of results is known to be highly controversial, and in gravity it is even more controversial than in electrodynamics. For example, in [86] it was claimed that, if one suggests that the b_μ is the vacuum expectation value (VEV) of a dynamical field, the correct result for the four-dimensional gravitational CS term is zero, as is also required by the gauge invariance of the Lagrangian (and not only of the action). Nevertheless, the question whether the requirements of [86] are indeed so necessary is still open, as the presence of ambiguities in generic Lorentz-breaking theories is a strongly polemical problem.

However, there are also other interesting scalar-tensor gravity models which we will consider now.

3.3 Brans-Dicke Gravity

The Brans-Dicke (BD) gravity is one of the most known and studied scalar-tensor gravity models. Originally, it has been introduced in [87], basing on the idea that the physical space itself possesses geometrical features beyond those generated by matter (this is one of the forms of the so-called Mach principle), so, the action of the BD gravity was proposed in the form

$$S = \int d^4x \sqrt{|g|} \left(\phi R - \frac{\omega}{\phi} \partial_\mu \phi \partial^\mu \phi + 16\pi \mathcal{L}_{mat} \right). \tag{3.36}$$

In this theory, the new scalar field ϕ (which does not contribute to the matter Lagrangian) plays the role of the effective gravitational constant; indeed, if one chooses $\phi = \frac{1}{2k^2}$, the theory reduces to the Einstein gravity with the usual matter. One advantage of the theory consists in the fact that the coupling constant ω is dimensionless, hence the negative-dimension constants jeopardizing the renormalizability of the gravity are ruled out. Also, in this case the gravitational constant has a dynamic origin being related with an asymptotic value of the ϕ.

For this theory, one can derive equations of motion:

$$\frac{2\omega}{\phi} \Box \phi - \frac{\omega}{\phi^2} \partial_\mu \phi \partial^\mu \phi + R = 0; \tag{3.37}$$

$$R_{\mu\nu} - \frac{1}{2} g_{\mu\nu} R = \left(\frac{8\pi}{\phi} \right) T_{\mu\nu} + \frac{\omega}{\phi^2} \left(\partial_\mu \phi \partial_\nu \phi - \frac{1}{2} g_{\mu\nu} \partial_\rho \phi \partial^\rho \phi \right) + \frac{1}{\phi} \left[\nabla_\nu (\partial_\mu \phi) - g_{\mu\nu} \Box \phi \right],$$

where $T_{\mu\nu}$ is the energy-momentum tensor of the usual matter (not including ϕ). Contracting this equation with $g^{\mu\nu}$, we find

$$R = - \left(\frac{8\pi}{\phi} \right) T + \frac{\omega}{\phi^2} \partial_\rho \phi \partial^\rho \phi + \frac{3}{\phi} \Box \phi, \tag{3.38}$$

which we can combine with Eq. (3.37), obtaining

$$\Box \phi = \left(\frac{8\pi}{3 + 2\omega} \right) T. \tag{3.39}$$

Equations (3.38, 3.39) are analogues of the Einstein equations and can be solved.

As a first example, we consider the static spherically symmetric metric (3.20) which we now rewrite as

$$ds^2 = e^{2\alpha(r)} dt^2 - e^{2\beta(r)} (dr^2 + r^2(d\theta^2 + \sin^2\theta d\phi^2)). \tag{3.40}$$

In the vacuum case, $T_{\mu\nu} = 0$, this metric will be a consistent solution of equations of motion [87]. Explicitly, one finds

$$e^{\alpha(r)} = e^{\alpha_0} \left[\frac{1 - \frac{2B}{r}}{1 + \frac{2B}{r}} \right]^{1/\lambda};$$

$$e^{\beta(r)} = e^{\beta_0} (1 + \frac{2B}{r})^2 \left[\frac{1 - \frac{2B}{r}}{1 + \frac{2B}{r}} \right]^{(\lambda - C - 1)/\lambda}; \qquad (3.41)$$

$$\phi(r) = \phi_0 e^{\alpha_0 C} \left[\frac{1 - \frac{2B}{r}}{1 + \frac{2B}{r}} \right]^{C/\lambda},$$

with $\alpha_0, \beta_0, B, C, \lambda$ being some constants.

The cosmological solutions also were found in [87], and in the vacuum case, they look like

$$\phi = \phi_0 t^r, \quad a = a_0 t^q;$$
$$r = \frac{2}{4 + 3\omega}, \quad q = \frac{2 + 2\omega}{4 + 3\omega}, \qquad (3.42)$$

so, accelerating solutions ($q > 1$) are possible for $\omega < -2$. Further, various papers, continuing this study, discussed cosmic acceleration in BD gravity in detail; see, for example, [88].

Now, let us discuss the Gödel-type solutions (1.10) in the BD gravity. It has been shown in [89] that the nontrivial solution, i.e. the one with a non-constant scalar ϕ (otherwise the BD gravity reduces trivially to the Einstein gravity) is possible only if the action (3.36) includes the cosmological constant as well, so, one has

$$S = \int d^4x \sqrt{|g|} \left(\phi(R - 2\Lambda) + \frac{\omega}{\phi} \partial_\mu \phi \partial^\mu \phi + 16\pi \mathcal{L}_{mat} \right). \qquad (3.43)$$

The modified Einstein equations, in the tetrad base, look like

$$G_B^A - \delta_B^A \Lambda = \left(\frac{8\pi}{\phi} \right) T_B^A - \frac{\omega}{\phi^2} \left(\partial^A \phi \partial_B \phi - \frac{1}{2} \delta_B^A \partial_C \phi \partial^C \phi \right) +$$
$$+ \phi^{-1} \left(\nabla_B \partial^A \phi - \delta_B^A \Box \phi \right), \qquad (3.44)$$

with the equation of motion for ϕ again given by (3.39). Choosing again the matter in the form of a composition of the fluid and electromagnetic field (see Sect. 3.2.2), with the angular velocity parametrizing the Gödel-type metric (1.10), we find that the case $\phi = \phi(z)$ yields

$$4\Omega^2 - m^2 = \left(\frac{8\pi}{\phi} \right) (\rho + E_0^2), \quad m^2 + 2\Lambda = -\frac{\phi''}{\phi}, \qquad (3.45)$$

with $\phi'' = -\frac{1}{3 + 2\omega} (8\pi \rho + 2\Lambda \phi)$ being the equation of motion for the scalar field.

The typical cases of solutions are

(i) $4\Omega^2 - m^2 = 0$ (that is, the RT causal solution [5]), $\rho = E_0 = 0$. In this case ϕ yields an exponential (hyperbolic or trigonometric) solution.
(ii) $\rho = const$, $\phi = const$—trivial case reducing to GR.

For $\phi = \phi(t)$, one arrives at $\phi = const$, and this case is again trivial. In principle, more involved situations can be studied as well. As for the black hole solutions in BD gravity, we strongly recommend the classical paper [90]. In principle, many other solutions for the BD gravity have been studied, including global monopoles, wormholes, etc. but the restricted volume of this book does not allow for their detailed discussion.

3.4 Galileons

One of the most important examples of the scalar-tensor gravity models is the Galileons theory (also known as Horndeski theory) proposed originally in [91]. Its key idea is as follows: let us consider the most general scalar-tensor action which involves no more than second derivatives of the metric tensor and no more than the first ones of the scalar field. Effectively, it was a suggestion of the Lovelock-like construction not only in the geometric sector but also in the scalar one. So, we suggest the action to look like

$$S = \int d^4x \sqrt{|g|} \mathcal{L}(g_{\mu\nu}, \partial_\lambda g_{\mu\nu}, \partial_\lambda \partial_\rho g_{\mu\nu}; \phi, \partial_\mu \phi). \qquad (3.46)$$

As a result, the equations of motion include various tensors constructed on the base of the Riemann curvature and its covariant derivatives, and various derivatives of the scalar field. In principle, we can have the equations of motion for the gravitational field with Lovelock-like l.h.s. and non-canonical scalar-dependent r.h.s., and strongly nonlinear equations of motion for the scalar. We note that there is no ghost problem here since there is no higher derivatives. In principle, even on the flat background, one can have a theory of a scalar field with highly nonlinear equation of motion, the so-called K-theory (see [92] and references therein).

However, the model (3.46) was forgotten for a long time and revitalized only in 2008, in the paper [93] where the concept of galileons was formulated. Its key idea consists in the invariance of the theory with respect to the combination of dilatations and conformal transformations so that the new scalar π transforms as $\pi \to \pi + c + b_\mu x^\mu$, where c and b_μ are constants. These transformations look similar to the Galilean ones, therefore the π was called the galileon. So, again, the pivotal idea is that we have derivative couplings but no higher derivatives in the kinetic term.

There are five terms with the symmetry above. Let us introduce notations $\Pi^{\mu\nu} = \partial^\mu \partial^\nu \pi$, $[A] = A^\mu_\mu$ for trace (so, $\frac{1}{2}[\Pi]\partial\pi \cdot \partial\pi = \frac{1}{2}\Box\pi \partial^\mu \pi \partial_\mu \pi$), $[\Pi] = \Box\pi$, etc. and use a dot for the usual scalar product like $A \cdot B \equiv A_\mu B^\mu$. So, we can write our five terms as

$$\mathcal{L}_1 = \pi,$$

$$\mathcal{L}_2 = -\frac{1}{2}\partial\pi \cdot \partial\pi;$$

$$\mathcal{L}_3 = -\frac{1}{2}[\Pi]\partial\pi \cdot \partial\pi;$$

$$\mathcal{L}_4 = -\frac{1}{4}\Big([\Pi]^2\partial\pi \cdot \partial\pi - 2[\Pi]\partial\pi \cdot \Pi \cdot \partial\pi - [\Pi^2]\partial\pi \cdot \partial\pi + 2\partial\pi \cdot \Pi^2 \cdot \partial\pi\Big);$$

$$\mathcal{L}_5 = -\frac{1}{5}\Big([\Pi]^3\partial\pi \cdot \partial\pi - 3[\Pi]^2\partial\pi \cdot \Pi \cdot \partial\pi - 3[\Pi][\Pi^2]\partial\pi \cdot \partial\pi$$
$$+6[\Pi]\partial\pi \cdot \Pi^2 \cdot \partial\pi + 2[\Pi]^3\partial\pi \cdot \partial\pi$$
$$+3[\Pi^2]\partial\pi \cdot \Pi \cdot \partial\pi - 6\partial\pi \cdot \Pi^3 \cdot \partial\pi\Big). \tag{3.47}$$

The complete Lagrangian of π is a linear combination of these terms: $\mathcal{L} = c_1\mathcal{L}_1 + c_2\mathcal{L}_2 + c_3\mathcal{L}_3 + c_4\mathcal{L}_4 + c_5\mathcal{L}_5$. Clearly, the next step consists in the coupling of these Lagrangians to gravity, but let us first describe some perturbative effects of these couplings.

One of the interesting effects is that these galileon terms $\mathcal{L}_i$ are not renormalized by quantum corrections! The reasons are as follows [94]. First, the galileon is massless, so, its propagator is $1/k^2$. Then, all galileon couplings c_3, c_4, c_5 have negative mass dimensions, therefore the quantum contributions to these terms possess quadratic and even higher divergences. After integration of subloops, the leading divergence is proportional to $\int d^4k(k^2)^n$, with $n \geq -1$, and this integral vanishes within dimensional regularization. Finally, the subleading contributions to galileon vertices vanish as well (this proof is more sophisticated being based on the analysis of symmetries; see [94]). In principle, such conclusions are natural for a massless theory with derivative couplings. Other divergent contributions to the effective action in the galileons theory in the flat space, which do not match the form of the classical action, in particular, involve more derivatives (for example, $\Box^2$ terms), are discussed in [95].

The key point in studying galileons is constructing their interaction with gravity. In the simplest case, they are coupled to the curvature (or, as is the same, to the Einstein tensor $G_{\mu\nu}$), so we have terms like [96, 97]

$$\delta S_4 = \int d^4x\sqrt{|g|}(\pi_\mu\pi^\mu)(\pi_\nu G^{\nu\rho}\pi_\rho), \tag{3.48}$$

with $\pi_\mu \equiv \nabla_\mu\pi$, etc. or the higher terms like $\pi_\mu\pi^{\mu\nu}\pi^\rho G_{\nu\rho}$, or the simplest term $\pi^\mu\pi^\nu G_{\mu\nu}$ (the last term is the example of the John term; see below). So, effectively we have a gravity-coupled scalar field with strongly nonlinear dynamics involving interaction vertices including derivatives of fields. As has been claimed in [97], these terms are of special interest within the cosmological context, where it has been explicitly shown that the solutions with constant $H = \frac{\dot{a}}{a}$ are consistent for the presence of galileons, therefore de Sitter-like exponential expansion is possible in this case, with neither potential term for the scalar nor cosmological constant being employed; therefore the galileons theory is a sound candidate for the role of the dark energy. In [98], it was argued that only the minimal scalar-gravity couplings must be considered; as a result, there were introduced four typical galileon-gravity coupling terms called John, Paul, George and Ringo:

$$
\begin{aligned}
\mathcal{L}_{John} &= V_J(\pi) G_{\mu\nu} \nabla^\mu \pi \nabla^\nu \pi; \\
\mathcal{L}_{Paul} &= V_P(\pi) P_{\mu\nu\rho\sigma} \nabla^\mu \pi \nabla^\nu \pi \nabla^\rho \nabla^\sigma \pi; \\
\mathcal{L}_{George} &= V_G(\pi) R; \\
\mathcal{L}_{Ringo} &= V_R(\pi) \mathcal{G},
\end{aligned}
\tag{3.49}
$$

where $P^{\mu\nu\alpha\beta} = -\frac{1}{4} \epsilon^{\mu\nu\rho\sigma} \epsilon^{\alpha\beta\gamma\delta} R_{\rho\sigma\gamma\delta}$ is the double dual of the Riemann curvature. In [98], the cosmological aspects of the theory involving these terms were studied; especially, it was explained how the known cosmological self-tuning problem is solved in this theory. Various issues related to the cosmic acceleration in the presence of galileons are studied numerically also in [99]. Many other papers are also devoted to galileon cosmology. As a result, one can conclude that the galileon field is a sound candidate for a role of the dark energy.

Nevertheless, while the galileons are mostly considered within the cosmological framework, there are some papers on their applications within contexts of other solutions. For example, consistency of various types of black holes, including regular ones, and of other static spherically symmetric solutions within the quadratic Horndeski theory, i.e. when only $\mathcal{L}_2$ presented in (3.47) is taken into account, was shown in [100–103] with the use of rather involved calculations and numerical studies. An interesting review of galileons is presented in [104].

The Gödel-type metric (1.10) also was treated within the galileon gravity. In this case, the action of galileons is given by the sum of first three terms of (3.47), $\mathcal{L}_1, \mathcal{L}_2$ and $\mathcal{L}_3$, multiplied by M^3, 1 and M^{-3} respectively in order to provide the correct mass dimensions of these terms, so that M is a constant with mass dimension 1, were taken into account [105], so that the galileon field π yields an additional contribution to the energy-momentum tensor. Straightforward calculations done in [105] show that, for the case when π depends only on the zero coordinate t, both Gödel and Gödel-type metrics solve the modified Einstein equations only in a trivial case $\pi = const$, so that the situation reduces to the GR.

The situation becomes less trivial if we, for the cylindrically symmetric Gödel-type metric (1.10), assume the galileon field to depend only on the coordinate along the symmetry axis, i.e. $\pi = \pi(z)$. In this case, the modified Einstein equations, in the tetrad basis, look like

$$
G_{AB} + \Lambda g_{AB} = \kappa^2 \left[T_{AB}^{(m)} + T_{AB}^{(\pi)} \right],
\tag{3.50}
$$

i.e. the galileon contribution to the energy-momentum tensor is given by $T_{AB}^{(m)}$ whose nontrivial components are

$$T_{(0)(0)}^{(\pi)} = -\frac{c_1 M^3}{2}\pi - \frac{c_2}{2}\pi'^2 + \frac{c_3}{M^3}\pi'^2\pi'', \tag{3.51}$$

$$T_{(1)(1)}^{(\pi)} = T_{(2)(2)}^{(\pi)} = \frac{c_1 M^3}{2}\pi + \frac{c_2}{2}\pi'^2 - \frac{c_3}{M^3}\pi'^2\pi'', \tag{3.52}$$

$$T_{(3)(3)}^{(\pi)} = \frac{c_1 M^3}{2}\pi - \frac{c_2}{2}\pi'^2. \tag{3.53}$$

Here the prime symbol is for the derivative with respect to z. The galileon field satisfies the equation of motion $\pi'' = 0$. Thus, the non-zero components of the modified Einstein equations, in the presence of dust-like matter with density ρ and pressure p and electromagnetic field with the amplitude e (i.e. the same variables used in Sect. 3.2.2), i.e. (00), (11) and (33) ones respectively, read as

$$3\Omega^2 - m^2 - \Lambda = \kappa^2\rho + \kappa^2\frac{e^2}{2}$$
$$+ \kappa^2\left[-\frac{c_1 M^3}{2}\pi - \frac{c_2}{2}\pi'^2 + \frac{c_3}{M^3}\pi'^2\pi''\right], \tag{3.54}$$

$$\Omega^2 + \Lambda = \kappa^2 p + \kappa^2\frac{e^2}{2} + \kappa^2\left[\frac{c_1 M^3}{2}\pi + \frac{c_2}{2}\pi'^2 - \frac{c_3}{M^3}\pi'^2\pi''\right], \tag{3.55}$$

$$m^2 - \Omega^2 + \Lambda = \kappa^2 p - \kappa^2\frac{e^2}{2} + \kappa^2\left[\frac{c_1 M^3}{2}\pi - \frac{c_2}{2}\pi'^2\right]. \tag{3.56}$$

Straightforward analysis of these equations, performed in [105], shows that various situations, generating different relations between constant parameters of the theory, are possible. For example, at $c_1 = 0$ (we note that this condition is rather reasonable since it allows to rule out the term which is linear in π field, and such terms are usually avoided within quantum field theory context), the causal Gödel-type solutions are possible, for example, at $e = \rho = p = 0$ and $\pi(z) = B + Az$ (i.e. we have vacuum, and the galileon is treated as an ingredient of a full description of the gravity), the causal RT solution [5] with $m^2 = 4\Omega^2$ is possible. We note that this form of the solution for the galileon field resembles the solution for the CS coefficient found in Sect. 3.2.2. Other interesting solutions can be found as well [105].

To close this section, we note that many aspects of galileons still must be studied.

3.5 Conclusions

We formulated several examples of scalar-tensor gravity models whose form does not match the standard quintessence-gravity Lagrangian $\mathcal{L} = \sqrt{|g|}(\frac{1}{2\kappa^2}R + \frac{1}{2}g^{\mu\nu}\partial_\mu \varphi\partial_\nu\varphi - V(\varphi))$ which is well studied, both within the cosmological and quantum field theory contexts (see, for example, [6, 7]). Explicitly, we considered the four-dimensional CS modified gravity, the Brans-Dicke gravity and the galileons theory. These theories display new interesting features.

First of all, the CSMG allows for the CPT symmetry breaking, and, for a certain form of the CS coefficient, also for the Lorentz symmetry breaking, opening thus a way for intensive studies of Lorentz-violating modifications of gravity. Some of these modifications will be discussed in the next chapter. Besides, in the presence of the gravitational CS term, new solutions, impossible within the usual GR, can arise.

Second, the Brans-Dicke gravity represents itself as a theory allowing to rule out the gravitational constant possessing negative mass dimension and hence implying problems with quantum description of the gravity. Moreover, it turns out that some new solutions, which are not consistent within the GR, are also possible in this case.

Third, the galileons theory turns out to be a sound candidate for a description of the dark energy allowing for accelerated solutions. Besides this, the galileon contributions to the action arise by applying the Stuckelberg approach for the massive gravity. Explicitly, at the first step one introduces the new vector field to construct the gauge invariant extension for the mass term of the gravity, and at the second step, to achieve the gauge symmetry for this vector field, one introduces the scalar field whose action matches the galileon form [52].

To conclude, for the scalar-tensor gravity models, one can arrive at essentially new results. One of the most interesting conclusions is the possibility to introduce the Lorentz symmetry breaking within the gravitational context, for a special form of the CS coefficient. However, it is clear that in this context, an extension of gravity through introduction of vector fields seems to be more promising since the vacuum expectations of vector fields can yield constant vectors necessary to introduce privileged space-time directions breaking thus the Lorentz symmetry.

Chapter 4
Vector-Tensor Gravities and Problem of Lorentz Symmetry Breaking in Gravity

4.1 Introduction and Motivations

The interest in vector-tensor gravity models strongly increased in recent years. One of the main motivations to studying these models arises from the idea of the Lorentz symmetry breaking. Indeed, as is well known, in the flat space the explicit Lorentz symmetry breaking is implemented through the introduction of a constant vector (tensor) generating a space-time anisotropy (see, for example, [106–108]). As we already noted in the previous chapter, this methodology allowed to define, for example, the CFJ term (3.7) as well as many other terms discussed in [106]. However, in the curved space the explicit Lorentz symmetry breaking faces serious problems. First of all, the definition of a constant vector (tensor) itself in this case becomes highly controversial: for example, while in the flat space the constant vector k^μ is defined to satisfy the condition $\partial_\nu k^\mu = 0$, this condition cannot be applied in a curved space since it breaks the general covariance. The possible "covariant extension" of this condition like $\nabla_\nu k^\mu = 0$ would imply extra restrictions for the space-time geometry referred in [109] as no-go constraints (and, moreover, nobody could guarantee these restrictions to be satisfied for a general choice of a vector k_μ as well as for other constant tensors). In principle, one can also deal with derivative expansions of corresponding effective actions, where various orders of derivatives of "constant" tensors can be obtained (see, for example, [110]); however, it is clear that in this case the definition of a constant vector (or tensor) simply loses its sense, and such a vector becomes an extra field. Moreover, in many cases such possible new terms are not gauge invariant which means that together with the Lorentz symmetry, the general covariance for such terms is broken as well (the problem of breaking the general covariance in modified gravity is discussed in detail in [111]; in principle, it should be noted that breaking of general covariance occurs for the term $u^\mu u^\nu R_{\mu\nu}$ proposed in [112] as a possible example of a CPT-even LV term for gravity, as well as for the one-derivative linearized LV term discussed in [77]).

Therefore, the most appropriate method for implementing the Lorentz symmetry breaking into a curved space-time turns out to be based on the spontaneous symmetry

A. Petrov et al., *Introduction to Modified Gravity*,
SpringerBriefs in Physics, https://doi.org/10.1007/978-3-031-46634-2_4

breaking. Its essence is as follows. One considers the action of the metric tensor coupled to the vector field (again, similar to the previous chapter, this vector field is treated as an ingredient of gravity model itself rather than the matter; thus, we have the vector-tensor gravity) so that the purely geometric sector is presented by the usual Einstein-Hilbert action, and the dynamics of the vector field is described by the Maxwell-like term, plus a potential whose minimum yields a vector implementing the Lorentz symmetry breaking, and maybe also some extra terms responsible for a vector-gravity coupling. The paradigmatic example is the bumblebee action [113] (the name "bumblebee" itself was introduced in [114]), looking like

$$ S = \int d^4x \sqrt{|g|} \left(\frac{1}{2\kappa^2}(R + \xi B^\mu B^\nu R_{\mu\nu}) - \frac{1}{4}B_{\mu\nu}B^{\mu\nu} - V(B^\mu B_\mu \mp b^2) \right) . (4.1) $$

Here ξ is a constant (chosen to be zero in the minimal case), $B_{\mu\nu} = \partial_\mu B_\nu - \partial_\nu B_\mu$ is the stress tensor for the bumblebee field B_μ and V is the potential possessing an infinite set of minima $B_{0\mu}$ satisfying the condition $B_0^\mu B_{0\mu} = \pm b^2$ (the difference of signs reflects that the vector $B_{0\mu}$ can be either time-like or space-like, while $b^2 > 0$). In principle, instead of the dimension-4 term $B^\mu B^\nu R_{\mu\nu}$, we can consider any LV terms listed in [109], with the LV vectors (tensors) replaced by corresponding products of various degrees of the bumblebee field B^μ; for example, we can use the dimension-3 term $B^\mu \Gamma^\alpha_{\mu\alpha}$, or the dimension-5 term $B^\mu K_\mu$, where K_μ is the Chern-Simons topological current (3.10), etc. So, actually choosing of one of the vacua $B_{0\mu}$ allows to introduce the privileged direction. The potential is usually chosen to be quartic in the field B_μ by renormalizability reasons. Alternatively, one can deal with Einstein-aether theory where, instead of this, the minima arise due to a constraint multiplied by a Lagrange multiplier σ, so that one has $V = \sigma(B^\mu B_\mu \mp b^2)$, but the kinetic term is not Maxwell-like being a more generic quadratic function of covariant derivatives of the vector B_μ. In principle, one can consider the vector-tensor gravity models without any potential [115, 116]; however, in this case the spontaneous Lorentz symmetry breaking cannot occur. Such theories are considered mostly within the cosmological context (see, for example, [115]).

Within this chapter, we discuss some interesting classical results for the Einstein-aether gravity and for the bumblebee gravity. At the end of the chapter, we also will review some terms proposed in [106, 108] as possible extensions of the Einstein gravity allowing to break the Lorentz symmetry explicitly. As for the Horava-Lifshitz gravity, although it represents itself as an example of non-Lorentz-invariant gravity model, it is described in terms of the essentially distinct methodology and will be discussed in the next chapter.

4.2   Einstein-Aether Gravity

So, let us implement the spontaneous Lorentz symmetry breaking in a curved space-time. To justify the importance of this approach, one can remind that namely the spontaneous breaking mechanism has been initially proposed to explain the origin of the Lorentz symmetry breaking in the low-energy limit of the string theory [117]. Following this concept, one considers a vector field B_μ with a constant square, i.e. $B^\mu B_\mu = \pm b^2$, which is implemented by introducing the constraint with the use of the Lagrange multiplier σ, adding to the Lagrangian the potential $V = \sigma(B^\mu B_\mu \mp b^2)$. Alternatively, as we already noted above, one can introduce the quartic potential. The approach based on the Lagrange multiplier has been adopted within the Einstein-aether gravity originally formulated in the paper [118]. In this case, the above constraint is generalized to a curved space-time as $g^{\mu\nu} u_\mu u_\nu - 1 = 0$, where u_μ is the so-called aether vector field introducing the privileged space-time frame and appears to be a reminiscence of the pre-Einsteinian aether concept, since the u^μ can be naturally interpreted as a 4-velocity of some fundamental frame which can be called the aether frame. Further, this vector was used in [112] to construct aether-like CPT-even Lorentz-violating terms involving various fields; in particular, in the gravity case the already mentioned $u^\mu u^\nu R_{\mu\nu}$ term was introduced.

Our starting point is the action [118]

$$S = -\frac{1}{2\kappa^2} \int d^4x \sqrt{|g|} \Big[R + \lambda(u^\mu u_\mu - 1) + K^{\alpha\beta}_{\mu\nu} \nabla_\alpha u^\mu \nabla_\beta u^\nu \Big], \qquad (4.2)$$

where

$$K^{\alpha\beta}_{\mu\nu} = c_1 g^{\alpha\beta} g_{\mu\nu} + c_2 \delta^\alpha_\mu \delta^\beta_\nu + c_3 \delta^\alpha_\nu \delta^\beta_\mu + c_4 u^\alpha u^\beta g_{\mu\nu}. \qquad (4.3)$$

This action involves an above-mentioned constraint introduced with the use of the Lagrange multiplier λ. The c_1, c_2, c_3, c_4 are some dimensionless constants. It is interesting to note that the term $R_{\alpha\beta} u^\alpha u^\beta$ proposed as the aether term in [112] arises in this theory (together with some other terms) for the particular case $c_3 = -c_2$ when the commutator of covariant derivatives yielding a curvature tensor emerges [119].

The corresponding equations of motion look like [119]:

$$g_{\alpha\beta} u^\alpha u^\beta = 1; \quad \nabla_\alpha J^\alpha_\mu - c_4 \dot{u}_\alpha \nabla_\mu u^\alpha = \lambda u_\mu;$$

$$R_{\alpha\beta} - \frac{1}{2} R g_{\alpha\beta} = T_{\alpha\beta} = -\frac{1}{2} g_{\alpha\beta} \mathcal{L}_u + \nabla_\mu \Big(J^\alpha_{(\mu} u_{\beta)} - J^\mu_{(\alpha} u_{\beta)} - J_{(\alpha\beta)} u^\mu \Big) + \qquad (4.4)$$

$$+ c_1[(\nabla_\mu u_\alpha)(\nabla^\mu u_\nu) - (\nabla_\alpha u_\mu)(\nabla_\beta u^\mu)] + c_4 \dot{u}_\alpha \dot{u}_\beta + [u_\nu \nabla_\mu J^{\mu\nu} - c_4 \dot{u}^2] u_\alpha u_\beta.$$

Here, $\dot{u}^\mu = u^\alpha \nabla_\alpha u^\mu$, $J^\alpha_\mu = K^{\alpha\beta}_{\mu\nu} \nabla_\beta u^\nu$ and $\mathcal{L}_u$ is the u-dependent part of the Lagrangian. We note again that the vector u_μ has nothing to do with the usual matter, being an ingredient of the full description of the gravity itself, so, the Einstein-aether theory is an example of the vector-tensor gravity.

So, now our task will consist in finding some solutions for these equations, or, to be more precise, in checking the consistency of known GR solutions within the Einstein-aether gravity.

As the simplest example, we choose the spherically symmetric static metric, which is consistent when the vector u_μ is chosen to be time-like, in order to satisfy the constraint. In our case, it is convenient to choose this metric in the form slightly different from (3.20), namely

$$ds^2 = N(r)dt^2 - B(r)(dr^2 + r^2(d\theta^2 + \sin^2\theta d\phi^2)). \tag{4.5}$$

The consistency of this metric within the Einstein-aether gravity has been verified within the perturbative methodology for various relations between the parameters c_1, c_2, c_3, for example one can choose the condition, $c_1 + c_2 + c_3 = 0$, and c_4 can be chosen to be zero without any problems since it can be removed through a simple change of variables (see details in [119]) so that the $N(r)$ and $B(r)$ turn out to be represented as power series in $x = 1/r$ providing that they tend to 1 at infinity as it must be, with some lower coefficients in these power series, up to $1/r^3$ terms in large r limit have been explicitly found in certain cases.

For example, treating the black hole solutions, one can show [119] that the metric

$$ds^2 = \left(1 - \frac{2M}{r} + \frac{2\beta M^2}{r^2}\right)dt^2 - \left(1 - \frac{2\gamma M}{r}\right)(dr^2 + r^2(d\theta^2 + \sin^2\theta d\phi^2)) \tag{4.6}$$

is consistent in this theory, with $\gamma = 1$, and β expressed in terms of coefficients c_1, c_2, c_3. This solution can be naturally treated as a modification of the Schwarzschild metric.

Similarly, much more solutions for the Einstein-aether gravity can be obtained, in particular, the cosmological ones. In this context, the detailed study of various cosmological aspects of this theory has been performed in [120, 121] where the model involving two scalar fields coupled to Einstein-aether gravity was considered, and it has been explicitly demonstrated on the base of the numerical analysis of solutions that the consistent potential for these fields is the exponential one, and the de Sitter-like solutions can arise both in the past (inflationary Universe) and in the future (de Sitter attractor). Earlier, the idea of using the Einstein-aether model in order to explain the cosmic acceleration has been discussed in [122]. All this allows to conclude that the Einstein-aether gravity can be considered as an acceptable solution of the dark energy problem. Besides this, a detailed discussion of various aspects of Einstein-aether gravity, including a discussion of plane wave solutions and observational constraints on parameters of the theory, can be found in [123]. Also, we note that the Einstein-aether gravity also displays some similarity to the Einstein-Maxwell theory; see [118].

However, it is clear that the Einstein-aether model is problematic from the quantum viewpoint. Indeed, its action involves a constraint. As is well known (see, for example, [124]), a theory with constraints, being considered at the perturbative level, requires special methodologies like $1/N$ expansion which clearly cannot be applied to the Einstein-aether gravity since it involves only four fields u^μ. Moreover, in principle such a theory, when treated in an improper manner, can display various instabilities. Therefore, the natural idea consists in introducing the spontaneous Lorentz symmetry breaking not by constraints but by introducing some potential of the B_μ field displaying a set of minima. This idea gave origin to the bumblebee gravity [113, 114] which we begin to discuss now.

4.3 Bumblebee Gravity

So, let us start with considering the bumblebee gravity. Our initial point will be the action (4.1). The key features of this action, in comparison with the Einstein-aether theory, are the following ones.

First, this action is characterized by the generic potential, instead of the constraint, which makes it better for quantum studies since the usual perturbative methodology can be applied. Second, the kinetic term is Maxwell-like which is essential to avoid the arising of ghost modes [125]. Again, the $\pm$ sign reflects the fact that $b^2 > 0$. We note again that the vacua $B_{0\mu}$ are given by the condition $B_0^\mu B_{0\mu} = \pm b^2$, and these vacua are not required to be constants, in a curved space-time, which avoids the difficulties connected with the definition of the constant vectors in this case.

The first effect to note here is that after Lorentz symmetry breaking, we will have Nambu-Goldstone modes: if we introduce the vector b_μ corresponding to one of the vacua, i.e. $b^\mu b_\mu = \pm b^2$, define $B_\mu = b_\mu + A_\mu$ and rewrite the action (4.1) in terms of b_μ and A_μ, the resulting form of the action will be given by the Maxwell term, plus the axial gauge term proportional to $(b^\mu A_\mu)^2$, plus new couplings of the vector A_μ with the curvature, like $A^\mu A^\nu R_{\mu\nu}$, plus the aether-like term $b^\mu b^\nu R_{\mu\nu}$ [112].

Let us discuss some exact solutions for this theory. First, we consider the static spherically symmetric metric, following the lines of [114]. For the reasons of convenience, we rewrite the metric (3.20) as

$$ds^2 = -e^{2\varphi(r)}dt^2 + e^{2\rho(r)}dr^2 + r^2(d\theta^2 + \sin^2\theta d\phi^2). \quad (4.7)$$

Then, we choose the vacuum vector to be purely radial, i.e. $b_\mu = (0, b(r), 0, 0)$; thus one has $\nabla_\mu b_\nu = 0$ if $b(r) = \xi^{-1/2} b_0 e^{\rho(r)}$, with ξ being a constant.

For this metric, we find the only non-zero component of the Ricci tensor and the corresponding scalar curvature to be

$$R_{rr} = \frac{2\rho'}{r}; \quad R = \frac{2[1 + 2(r\rho' - 1)e^{-2\rho}]}{r^2}. \quad (4.8)$$

It is convenient to introduce a new dynamical variable $\Psi = \frac{1-e^{-2\rho}}{r^2}$. Its action will look like

$$S = \frac{2}{\kappa^2} \int dtdr r^2 e^{\rho+\varphi}[(3 + b_0^2)\Psi + (1 + \frac{b_0^2}{2} r\Psi')], \quad (4.9)$$

where b_0 was defined above.

The equation of motion, after varying it with respect to ϕ, is

$$(3 + b_0^2)\Psi + \left(1 + \frac{b_0^2}{2}r\Psi'\right) = 0. \qquad (4.10)$$

Its solution is $\Psi(r) = \Psi_0 r^{L-3}$, with $3 - L = (3 + b_0^2)/(1 + b_0^2/2)$, and

$$g_{rr} = e^{2\rho} = (1 - \Psi_0 r^{L-1})^{-1}, \qquad (4.11)$$

so, this component is similar to g_{rr} of the Schwarzschild metric, therefore our solution is characterized by the event horizon. In principle, more results for this metric can be obtained, for example, the Hawking temperature [114]. The case when the b_μ vacuum vector possesses not only the radial component but also the temporal one has been also discussed in [114]; as a result, the Schwarzschild-like solution will carry an extra factor $e^{\pm 2K_i r^\alpha}$, where α is a constant, the sign $+$ is for the temporal component and the sign $-$ for the radial one, with the values of K_i being different for these two components. Therefore, we conclude that the Lorentz symmetry breaking generates the black hole solutions. Other examples of black hole solutions in the bumblebee gravity, including Schwarzschild-(anti)de Sitter ones, were obtained and studied numerically in [126].

Another important example is the cosmological FRW metric. Here, we review its description within the bumblebee context presented in [127]. Explicitly, as a first attempt, we suggest the vector B_μ to be directed along the time axis, $B_\mu = (B(t), 0, 0, 0)$. Evidently, in this case the stress tensor for the bumblebee field vanishes, and the only nontrivial component of the equations of motion for the B_μ is

$$\left(V' - \frac{3\xi}{2\kappa^2}\frac{\ddot{a}}{a}\right)B = 0. \qquad (4.12)$$

Thus, the bumblebee field either vanishes or, at $\xi = 0$, stays at one of the minima of the potential. In this case, it is possible to show numerically that one has the de Sitter-like expansion of the Universe.

More generic cosmological solutions can be obtained for $B_{\mu\nu} \neq 0$. However, in this case the numerical analysis is necessary. Explicit studies carried out in [127] show that in this case, de Sitter-like solutions arise for many values of parameters of the theory confirming this a possibility to have a cosmic acceleration due to the bumblebee field; therefore, one can conclude that the spontaneous Lorentz symmetry breaking can explain the dark energy problem. In [128], the mechanism for arising cosmological anisotropies stemming from the bumblebee field is presented.

Finally, we consider also the Gödel solution (1.8). Within the bumblebee context, in the minimal case $\xi = 0$, it has been studied in [129]. In this case, the energy-momentum tensor is suggested to be a sum of the one for the relativistic fluid (we note that namely such a form has been employed in [4]):

$$T_{\mu\nu}^M = \rho v_\mu v_\nu + \Lambda g_{\mu\nu}, \tag{4.13}$$

and the one for the bumblebee:

$$T_{\mu\nu}^B = B_{\mu\alpha} B_\nu{}^\alpha - \frac{1}{4} g_{\mu\nu} B_{\lambda\rho} B^{\lambda\rho} - V g_{\mu\nu} + 2V' B_\mu B_\nu, \tag{4.14}$$

where V' is a derivative of the potential with respect to its argument, that is, $B^\mu B_\mu \mp b^2$. Therefore, the modified Einstein equation (in an appropriate system of units where $\kappa = 1$) looks like

$$G_{\mu\nu} = T_{\mu\nu}^M + T_{\mu\nu}^B. \tag{4.15}$$

The Einstein tensor $G_{\mu\nu}$ and the matter energy-momentum tensor $T_{\mu\nu}^M$ (4.13) in the bumblebee gravity are the same as in the usual Einstein gravity with the cosmological term. Therefore, the Gödel metric continues to be the solution in our theory if and only if the energy-momentum tensor of the bumblebee field vanishes. To achieve this situation, we suggest that the field B_μ is one of the vacua which, for the quartic potential $V = \frac{\lambda}{2}(B^\mu B_\mu \mp b^2)^2$, will yield vanishing of the potential and its derivative. So, it remains to find the vacuum for which the stress tensor $B_{\mu\nu} = \partial_\mu B_\nu - \partial_\nu B_\mu$ would vanish as well (the part proportional to Christoffel symbols vanishes identically). It is clear that the case of the constant B_μ is an excellent example. Some interesting cases of such vacua, for the metric in the form (1.8), are $B_\mu = (ab, 0, 0, 0)$, $B_\mu = (0, ab, 0, 0)$ and $B_\mu = (0, 0, 0, ab)$ (we note that the Gödel metric is characterized by the constant parameter a; cf. [4]).

It remains to check the consistency of these solutions with the equation of motion for the bumblebee field:

$$\nabla_\mu B^{\mu\nu} = 2V'(B^\mu B_\mu \mp b^2) B^\nu. \tag{4.16}$$

These equations are satisfied immediately. Indeed, the l.h.s. is zero since $B^{\mu\nu} = 0$ for these solutions, and its covariant derivative is also zero, and the r.h.s. is zero for the quartic potential, if B_μ is one of the vacua. Therefore, we conclude that the Gödel solution is consistent in the bumblebee gravity. More detailed discussion of this solution can be found in [129]. It is clear that a more generic Gödel-type solution (1.10) can be analyzed along the same lines. This study has been performed in [130], where it was shown that, to have a causal solution, one must require $\xi \neq 0$.

An interesting discussion of the bumblebee gravity is presented also in [131]. The starting point is the generalized bumblebee Lagrangian

$$\mathcal{L} = R - \zeta \bar{g}^{\alpha\gamma} \bar{g}^{\beta\delta} B_{\alpha\beta} B_{\gamma\delta} - V[B], \tag{4.17}$$

where $V[B]$ is a potential of the bumblebee field, explicitly, a function of the combination $B^\mu B_\mu \mp b^2$, with ζ being the coupling constant, and $\bar{g}^{\alpha\gamma} = g^{\alpha\gamma} + \beta B^\alpha B^\gamma$ the effective metric.

Then, we carry out background-quantum splitting for gravitational and bumblebee fields by the formulas $g_{\alpha\beta} = \eta_{\alpha\beta} + h_{\alpha\beta}$ and $B_\alpha = \bar{B}_\alpha + A_\alpha$, where $\bar{B}_\alpha$ is one of the vacua, i.e. $V|_{B=\bar{B}} = V'|_{B=\bar{B}} = 0$ (we note that, typically, the second derivative of the potential does not vanish at the vacuum, $V''|_{B=\bar{B}} \neq 0$!).

As a result, we arrive at the linearized equations of motion for the fluctuations $h_{\alpha\beta}$, A_α:

$$G_{\alpha\beta}| = V''|_{B=\bar{B}} \, \bar{B}_\alpha \bar{B}_\beta B^2|,$$

$$\bar{\eta}^{\alpha\delta} \bar{\eta}^{\beta\gamma} \partial_\beta F_{\gamma\delta}[A] = \frac{1}{2\zeta} V''|_{B=\bar{B}} \, \bar{B}^\alpha B^2|, \qquad (4.18)$$

where $|$ symbol is for a part linear in fluctuations $h_{\alpha\beta}$, A_α of our fields around backgrounds $\bar{g}_{\alpha\beta}$, $\bar{B}_\alpha$, for example, $B^2| = 2\bar{B}^\alpha A_\alpha - \bar{B}^\alpha \bar{B}^\beta h_{\alpha\beta}$, and $\bar{\eta}^{\alpha\delta} = \eta^{\alpha\delta} + \beta \bar{B}^\alpha \bar{B}^\delta$ (it is interesting to note that a similar metric arises within some studies of aether-like LV theories). The $F_{\gamma\delta}[A] = \partial_\gamma A_\delta - \partial_\delta A_\gamma$, as usual.

We can introduce background-dependent densities

$$\rho_m = -V''|_{B=\bar{B}} \, \bar{B}^2 B^2|,$$

$$\rho_e = \pm \frac{V''|_{B=\bar{B}} \sqrt{|\bar{B}^2|}}{2\zeta} B^2| \qquad (4.19)$$

and a 4-velocity $u_\alpha = \pm \dfrac{\bar{B}_\alpha}{\sqrt{|\bar{B}^2|}}$; as a result, the equations of motion become

$$G_{\alpha\beta}| = \rho_m u_\alpha u_\beta,$$

$$\partial^\beta F_{\beta\alpha}[A] = \rho_e u_\alpha, \qquad (4.20)$$

formally replaying the Einstein and Maxwell equations, respectively. Effectively, this means that our background field $\bar{B}_\mu$ plays the role of the charged dust. We note that, in principle, the $\bar{B}_\alpha$ and A_α fields can be coupled to usual matter in various manners being treated either as a usual photon or as a extra particle.

To conclude, we see that the bumblebee gravity can be treated as a sound candidate, first, to implement the Lorentz symmetry breaking within the gravity context, and, second, to display consistency with astronomical observations, due to the validity of most important general relativity solutions. Among other results, one can mention the study of dispersion relations in a linearized bumblebee gravity where the constant bumblebee field triggers deviations from the standard dispersion relations [132]. However, much more aspects of the bumblebee gravity, especially problem of validity and consistency of many other solutions, are still to be studied. Besides this, one of the most important issues to be treated in the future is the detailed investigation of perturbative aspects of the bumblebee gravity (some first studies in this direction, within the metric-affine geometry framework, are discussed in Chap. 7 of this book).

4.4 Comments on Linearized Lorentz-Breaking Terms in Gravity

Studies of weak gravity are very important within various contexts, such as gravitational waves, degrees of freedom and post-Newtonian approximation. As a result, the natural problem is the study of possible impacts of Lorentz symmetry breaking for the weak gravity case, that is, for the free (quadratic) action of the metric fluctuation $h_{\mu\nu}$. Effectively, we consider a theory given by the sum of the Fierz-Pauli Lagrangian (1.13) and possible additive Lorentz-breaking terms. The simplest example of such a term is the four-dimensional linearized CS term (3.9) discussed in Sect. 3.2. However, other LV extensions of linearized gravity can be introduced as well. As examples, one can consider weak field limits of all terms listed in [109].

Let us consider some typical cases of linearized LV additive terms in gravity (see [133] for details). The simplest LV additive term presented in [109] is the dimension-3 term $\mathcal{L}_g^{(3)} = \left(k_\Gamma^{(3)}\right)^\mu \Gamma_{\mu\alpha}^\alpha$. Denoting $\left(k_\Gamma^{(3)}\right)^\mu \equiv k^\mu$, for the weak gravity approximation, and taking into account that only transverse-traceless parts of metric fluctuations yield physical dynamics (see the discussion in [68]), we have $\sqrt{|g|} \simeq 1$, and can write $\mathcal{L}^{(3)} = -\frac{1}{2} k^\mu h^{\alpha\gamma} \partial_\mu h_{\alpha\gamma}$. This term, evidently, is nothing more as a total derivative, so, its contribution to dispersion relations (as well as other physical impacts) is trivial.

Another dimension-3 LV term possible in linearized gravity appears to have no analogue in full-fledged gravity, i.e. it cannot be expressed in terms of connections or Riemann (and hence Ricci) curvature tensors. It has been introduced in [77] and looks like

$$\mathcal{L}_{1d}^{(3)} = \epsilon^{\mu\nu\lambda\rho} b_\mu h_{\nu\alpha} \partial_\lambda h_\rho^\alpha. \tag{4.21}$$

This term has been briefly discussed in Sect. 3.2.3, where it was noted that in this case, the dispersion relations are nontrivially modified. We note that this term is not gauge invariant, which corresponds to breaking the general covariance; such a possibility has been discussed in [111].

The general form of the dimension-4 LV term is $\left(k_R^{(4)}\right)^{\mu\nu\lambda\rho} R_{\mu\nu\lambda\rho}$. In the particular case $\left(k_R^{(4)}\right)^{\mu\nu\lambda\rho} = b^\mu b^\lambda \eta^{\nu\rho} - b^\mu b^\rho \eta^{\nu\lambda} + b^\nu b^\rho \eta^{\mu\lambda} - b^\nu b^\lambda \eta^{\mu\rho}$, we arrive, up to overall numerical factor, at the aether-like expression for this term which takes the form $b^\mu b^\nu R_{\mu\nu}$ defined in [112]. The dispersion relations corresponding to the presence of this additive term have been studied in [132], where they were shown to look like $p^2 + \xi(b \cdot p)^2 = 0$, with ξ being some constant defined within the action of the model, and $b^2 p^2 - (b \cdot p)^2 = 0$. These relations are known to arise in certain non-gravitational LV theories as well [133].

The paradigmatic example of the dimension-5 LV term in gravity is the four-dimensional gravitational CS term, in the linearized case looking like (3.9), which has been discussed in great detail in Sect. 3.2 of this book. It was noted there that the dispersion relations in the gravity theory described by the sum of the Einstein-Hilbert and CS terms are the usual ones $E^2 = \mathbf{p}^2$; cf. [68]. Further, in [133] it was argued that apparently all higher derivative additive LV terms in gravity, if they have a generalized aether-like form being described by various degrees of only one LV vector b^μ, in certain cases will yield usual dispersion relations $E^2 = \mathbf{p}^2$. In [133], this statement was checked up to dimension-6 operators. However, it was shown that for the theory including both dimension-4 and dimension-6 operators, like in [134] (where the term $\kappa^{\mu\nu\lambda\rho} R_{\mu\nu\alpha\beta} R^{\alpha\beta}{}_{\lambda\rho}$ was considered), besides the standard dispersion relations, more complicated ones, like $\alpha + \beta p^2 + \gamma (b \cdot p)^2 = 0$, with α, β, γ being constant parameters of the theory, are present. We note, however, that for more involved LV additive terms in gravity, in particular, those characterized by more sophisticated constant LV tensors, the dispersion relations can display highly unusual forms; see the discussion in [135].

We note nevertheless that the linearized form of all terms listed in [109], except the gravitational CS term, breaks gauge symmetry. Nevertheless, several possibilities to formulate various gauge invariant additive LV terms for the metric fluctuation $h_{\mu\nu}$ were presented in [133]. Let us discuss some of these possibilities. Our starting point is the l.h.s. of the equations (1.14), that is, the linearized Einstein tensor $G^{(0)}_{\mu\nu}$. It is easy to check that $G^{(0)}_{\mu\nu}$ is gauge invariant. Hence, it is easy to construct gauge invariant LV additive terms to be added to the classical action, with use of of $G^{(0)}_{\mu\nu}$: the CPT-even one contains four derivatives and looks like

$$\mathcal{L}_{four} = \frac{1}{2} b^\mu G^{(0)}_{\mu\nu} b_\lambda G^{(0)\lambda\nu}. \tag{4.22}$$

The CPT-odd gauge invariant LV term, that can be constructed from $G^{(0)}_{\mu\nu}$, contains five derivatives:

$$\mathcal{L}_{five} = \frac{1}{2} \epsilon^{\alpha\beta\gamma\delta} b_\alpha b^\mu G^{(0)}_{\mu\beta} \partial_\gamma b^\nu G^{(0)}_{\nu\delta}. \tag{4.23}$$

We see that these terms essentially involve higher derivatives, whose presence is necessary to guarantee gauge invariance. It is clear that they can be easily extended for the case of presence of higher orders in derivatives, for example, by introducing additional degrees of the d'Alembertian operator.

In [133], also another manner to introduce gauge invariant LV terms in linearized gravity was proposed. Within it, we introduce the operator

$$\Pi^{\mu\nu} = \eta^{\mu\nu}\Box - \partial^\mu \partial^\nu, \tag{4.24}$$

which is actually the well-known transverse projector multiplied by the $\Box$ operator in order to avoid arising of nonlocal contributions. It satisfies the relation $\Pi^{\mu\nu}\Pi_{\nu\lambda} = \Box\Pi^\mu_\alpha$. Under usual linearized gauge transformations $\delta h_{\mu\nu} = \partial_\mu \xi_\nu + \partial_\nu \xi_\mu$, the object $\Pi^{\mu\nu} h_{\nu\alpha}$ is transformed as $\delta(\Pi^{\mu\nu} h_{\nu\alpha}) = \partial_\alpha \Box \xi^\mu - \partial_\alpha \partial^\mu (\partial \cdot \xi)$. So, we define the vector $K_\alpha = b_\mu \Pi^{\mu\nu} h_{\nu\alpha}$ transforming in this case as $\delta K_\alpha = \partial_\alpha [\Box(b \cdot \xi) - (b \cdot \partial)(\partial \cdot \xi)] = \partial_\alpha \Sigma[\xi]$, i.e. its variation is a gradient of some scalar, just as for the usual electromagnetic 4-potential A^μ. Therefore, we can construct gauge invariant objects with use of K_μ, so that they are similar to those for the electromagnetic field, that is,

$$\mathcal{L}_{even} = \frac{1}{2} K_\alpha \Pi^{\alpha\beta} K_\beta;$$
$$\mathcal{L}_{odd} = \epsilon^{\alpha\beta\gamma\delta} b_\alpha K_\beta \partial_\gamma K_\delta. \tag{4.25}$$

In the physical (transverse-traceless) sector, the CPT-odd term from (4.25) can be shown to be equivalent to the term (4.23). However, the disadvantage of these terms consists in the presence of higher derivatives which, as we noted already in Chap. 1, can generate ghost modes. It should be noted that, in Lorentz-breaking theories, specific forms of higher derivative terms can be shown to be free of higher time derivatives, and hence of ghosts [137]; for example, for the scalar field, the term $\frac{1}{\Lambda} \bar\Phi (n \cdot \partial)^3 \Phi$ does not involve higher time derivatives if the LV vector n_μ does not involve a nontrivial time component, $n^0 = 0$. However, while this mechanism, sometimes referred to as the Myers-Pospelov mechanism, has been already discussed in [137] for scalar, spinor and gauge fields, there are no known examples of its application to linearized gravity. Certainly, this issue must be studied.

4.5 Conclusions

We discussed vector-tensor gravity models. Just as in the previous chapter, the additional field, in this case the vector one, is treated not as a matter field but as an ingredient of the complete description of the gravity itself. The most important aspect of these models consists in the fact that some of them, namely those involving potential terms for the vector field, can be extremely useful within the context of the spontaneous Lorentz symmetry breaking. The known examples of these theories are the Einstein-aether gravity and the bumblebee gravity.

The Einstein-aether theory has been formulated earlier. Within it, the potential term generating the spontaneous Lorentz symmetry breaking is implemented through the constraint with the corresponding Lagrange multiplier field. From one side, this action is rather simple, but from another side, the presence of the constraint generates essential difficulties within the perturbative framework. Therefore, the bumblebee model is certainly much more promising. Moreover, the bumblebee approach displays an advantage in comparison with the naive application of the QFT approach suggesting to couple dynamical fields with constant vectors (tensors) which, as we already noted, cannot be consistently defined in a curved space-time.

The bumblebee approach allows to introduce many Lorentz-breaking vector-tensor terms. The term $B^\mu B^\nu R_{\mu\nu}$ from (4.1) is effectively nothing more that the gravitational aether term proposed in [112]. We note that treating of the B_μ as one of the bumblebee vacua rather than the usual constant vector allows to avoid breaking of the general covariance. In a similar manner, other LV gravitational terms introduced in [108] can be treated. As a result, relaxing the condition for the Lorentz-breaking vector to be constant, we have a theory consistent with the general covariance requirement.

We note that the term $B^\mu B^\nu R_{\mu\nu}$ is the particular case of the LV additive term $s^{\mu\nu} R_{\mu\nu}$ discussed in [108]. Actually, in [108], two LV terms are presented, so, the possible LV extension of gravity can be introduced by adding the term

$$S_{LV} = \int d^4x \sqrt{|g|}(s^{\mu\nu} R_{\mu\nu} + t^{\mu\nu\lambda\rho} R_{\mu\nu\lambda\rho}), \tag{4.26}$$

where $s^{\mu\nu}$, $t^{\mu\nu\lambda\rho}$ are coefficients of explicit Lorentz symmetry breaking (we note that, as we treat now the Riemannian formulation of gravity, we consider only the zero torsion case). However, up to now the main attention (see, for example, [132]) was paid to the $s^{\mu\nu}$ term while the $t^{\mu\nu\lambda\rho} = 0$ condition was applied. Nevertheless, it should be noted that the $t^{\mu\nu\alpha\beta}$ term actually does not yield any observable physical implications; this strange effect was called the t-puzzle [136], hence the assumption $t^{\mu\nu\lambda\rho} = 0$ appears to be very natural.

In principle, various additive LV terms proposed in [109] can be studied, including the already discussed terms, namely the one-derivative term $\left(k_\Gamma^{(3)}\right)^\mu \Gamma_{\mu\alpha}^\alpha$, the aether term $\left(k^{(4)}\right)^{\mu\nu\lambda\rho} R_{\mu\nu\lambda\rho}$, the CS term $\left(k_\mu^{(5)}\right) \epsilon^{\rho\mu\nu\lambda}(\Gamma_{\mu a}^{\ b}\partial_\nu\Gamma_{\lambda b}^{\ a} + \frac{2}{3}\Gamma_{\mu a}^{\ b}\Gamma_{\nu b}^{\ c}\Gamma_{\lambda c}^{\ a})$, and terms with two and more derivatives like $\kappa^{\mu\nu\lambda\rho} R_{\mu\nu\alpha\beta} R^{\alpha\beta}_{\ \ \lambda\rho}$, or those involving covariant derivatives of Riemann curvature tensor. Moreover, in principle, couplings of gravity not only with the vector bumblebee model but also with the tensor one discussed in [138] can be introduced. We note that, considering the dynamical antisymmetric tensor field $\tilde{B}^{\mu\nu}$, we can naturally define a new coupling $\tilde{B}^{\mu\nu} \tilde{B}^{\rho\sigma} R_{\mu\nu\rho\sigma}$ reproducing the last term in (4.26).

To close the discussion of the Lorentz symmetry breaking in gravity, let us say more words about the weak (linearized) gravity. We have noted already that, for the specific form of the CS coefficient, the gravitational CS term (3.6) displays Lorentz symmetry breaking. In [77], another, one-derivative LV term in the linearized gravity, the one given by (4.21), has been studied. In principle, much more Lorentz-breaking terms in the linearized gravity can be introduced. In this context, it is worth to mention also the massive LV term in gravity proposed in [58], looking like $\mathcal{L}_m = \frac{1}{8}(m_0^2 h_{00}^2 + 2m_1^2 h_{01}^2 - m_2^2 h_{ij}^2 + m_3^2 h_i^i h_j^j - 2m_4^2 h_0^0 h_i^i)$. For this term, the absence of vDVZ discontinuity was proved and some cosmological estimations were obtained in [58]. It is clear that a more generic massive term looking like $\frac{1}{2}\kappa^{\mu\nu\lambda\rho} h_{\mu\nu} h_{\lambda\rho}$, with constant $\kappa^{\mu\nu\lambda\rho}$, can be considered as well. However, it must be noted that many studies of Lorentz symmetry breaking in gravity are still to be carried out, and it is natural to expect that such studies be performed in the next years.

Chapter 5
Horava-Lifshitz Gravity

5.1 Introduction

As is well known, the most complicated problem of gravity is the problem of its consistent quantum description. Indeed, we have noted in Chap. 1 that the Einstein gravity is non-renormalizable since the mass dimension of the gravitational constant is negative. The natural improvement of a situation could consist in adding the higher derivative terms which clearly make the UV asymptotics of the propagator better. However, it is known that in this case the ghosts arise which makes the theory to be unstable; hence, higher derivative gravity models can be used only as effective theories for the low-energy domain.

Therefore, in [139], it was proposed to construct a desired extension of gravity, using only second time derivatives, so that the ghosts will be ruled out, thus the UV behavior of the propagator will be improved. Similar models for the scalar field, with modified kinetic terms like $\frac{1}{2}\phi(\partial_0^2 + (-1)^z \alpha \Delta^z)\phi$, have been introduced long ago within the condensed matter context in [140] where they were used to describe critical phenomena. In other words, we suggest that the Lorentz symmetry breaking is strong. Further, such theories with a strong difference between spatial and time directions have been denominated as theories with space-time anisotropy. The number z, defined in a manner similar to the action above (once more, if the action involves two time derivatives, it involves $2z$ spatial derivatives), is called the critical exponent. For the Lorentz-invariant theories, one has $z = 1$. To recover the Einstein limit, one must suppose that the action involves also lower derivative terms. One can verify that in such a theory, the dimension of the effective gravitational constant will depend on z, being actually equal to $z - d$, in a $(d + 1)$-dimensional space-time. Therefore, in $(3 + 1)$-dimensional space-time, the gravity model formulated on the base of the space-time anisotropy (further such theories became to be called the Horava-Lifshitz (HL) theories) is power-counting renormalizable at $z = 3$. However, it is clear that for such a theory, the perturbative calculations will be very involved.

In this chapter, we present a general review on HL gravity, introduce definitions of main quantities used within it, and describe most important classical solutions.

© The Author(s), under exclusive license to Springer Nature Switzerland AG 2023 57
A. Petrov et al., *Introduction to Modified Gravity*,
SpringerBriefs in Physics, https://doi.org/10.1007/978-3-031-46634-2_5

5.2 Basic Definitions

So, let us construct the gravity model on the base of a strong difference between time and space coordinates. Following the methodology developed in [139], we consider the space-time as a foliation $R \times M_3$, where R is the real axis corresponding to the time, and M_3 is the three-dimensional manifold parametrized by spatial coordinates. The most convenient variables to describe the gravitational field in this case are the Arnowitt-Deser-Misner (ADM) variables [141], that is, N, N_i, g_{ij} defined from the following representation of the metric (note that the signature of the metric in this chapter is $(-+++)$, different from the rest of the book):

$$
\begin{aligned}
ds^2 &= g_{\mu\nu}dx^\mu dx^\nu \equiv g_{00}\,dt^2 + 2\,g_{0i}\,dx^i dt + g_{ij}\,dx^i dx^j \\
&= -N^2 dt^2 + g_{ij}(dx^i + N^i dt)(dx^j + N^j dt),
\end{aligned}
\tag{5.1}
$$

so, g_{ij} is the purely spatial metric, and one can define the shift vector $N_i = g_{0i}$ and the lapse function $N = (g_{ij}N^i N^j - g_{00})^{1/2}$.

The Lagrangian was suggested to be in the form

$$
\begin{aligned}
L = \sqrt{g}N\Bigg(&\frac{2}{\kappa^2}(K_{ij}K^{ij} - \lambda K^2) - \frac{\kappa^2}{2w^4}C_{ij}C^{ij} + \frac{\kappa^2\mu}{2w^2}\frac{\epsilon^{ijk}}{\sqrt{g}}R_{il}\nabla_j R_k^l \\
&- \frac{\kappa^2\mu^2}{8}R_{ij}R^{ij} + \frac{\kappa^2\mu^2}{8(1-3\lambda)}\left[\frac{1-4\lambda}{4}R^2 + \Lambda R - 3\Lambda^2\right] + \mathcal{L}_m\Bigg),
\end{aligned}
\tag{5.2}
$$

where λ, w, μ are constants the R_{ij} is a purely spatial curvature constructed on the base of the spatial metric g_{ij}, and

$$
K_{ij} = \frac{1}{2N}(\dot{g}_{ij} - \nabla_i N_j - \nabla_j N_i)
\tag{5.3}
$$

is the extrinsic curvature, with the dot for a derivative with respect to t, $K = g^{ij}K_{ij}$, and

$$
C^{ij} = \frac{\epsilon^{ikl}}{\sqrt{g}}\nabla_k\left(R_l^j - \frac{1}{4}R\delta_l^j\right)
\tag{5.4}
$$

is a Cotton tensor. It involves three spatial derivatives; hence the term $C_{ij}C^{ij}$ is of the sixth order in derivatives. So, it is clear that the propagator in this theory behaves as $G(k) \sim \frac{1}{k_0^2 - k^6}$. As we already noted, this implies power-counting renormalizability of the theory, with the gravitational constant κ being indeed dimensionless.

The form of the Lagrangian (5.2) has been motivated by the "detailed balance" condition [139] requiring that the potential term (i.e. the part of the action which does not involve the extrinsic curvature K_{ij} which only includes the time derivatives) is

$$S_V = \frac{\kappa^2}{8} \sqrt{g} N \frac{\delta W}{\delta g_{ij}} G_{ijkl} \frac{\delta W}{\delta g_{kl}}, \tag{5.5}$$

where W is an action, and $G_{ijkl} = \frac{1}{2}(g_{ik}g_{jl} + g_{il}g_{jk} - \lambda g_{ij}g_{kl})$. For $z = 2$, one has $W = W_2 = \frac{1}{2\kappa_W} \int d^D x \sqrt{g}(R - 2\Lambda_W)$, and for $z = 3$, one chooses $W = W_3$ to be the three-dimensional Chern-Simons action, so, $\frac{\delta W_3}{\delta g_{ij}} = C^{ij}$ (a similar expression for the Cotton tensor in $2 + 1$ dimensions has been considered in Sect. 3.2), and substitution of $W = W_2 + W_3$ to (5.5) yields the potential term given by (5.2).

However, there are only very few attempts to do quantum calculations in the HL gravity see f.e. [142, 143]. Actually, in these papers the gravity is suggested to be a pure background field; only the matter is quantized. At the same time, it is clear that the calculations of quantum corrections in a pure HL gravity, besides being extremely involved technically, must answer the fundamental question—does the form of quantum corrections match the form of the classical action, i.e. is the HL gravity multiplicatively renormalizable? This question is still open.

Let us now write down the equations of motion for the HL gravity. We use the approach and notations from [144], with $Q_{kl} = N(\gamma R_{kl} + 2\beta C_{kl})$. It should be noted that the component g_{00} is not a fundamental dynamical variable of the theory being expressed in terms of N, N_i and g_{ij}. For g_{00}, one has

$$\frac{\delta S}{\delta g_{00}} = \left(\frac{\delta S_g}{\delta N} + \frac{\delta S_m}{\delta N} \right) \frac{\delta N}{\delta g_{00}} = G^{00} - T^{00} = 0. \tag{5.6}$$

We note that, since $N = (g_{ij}N^i N^j - g_{00})^{1/2}$, one has $\frac{\delta N}{\delta g_{00}} = -\frac{1}{2N}$. Hence,

$$G^{00} = \frac{1}{2N}(-\alpha(K_{ij}K^{ij} - \lambda K^2) + \beta C_{ij}C^{ij} + \sigma \tag{5.7}$$

$$+ \gamma \frac{\epsilon^{ijk}}{\sqrt{g}} R_{il} \nabla_j R_k^l + \zeta R_{ij} R^{ij} + \eta R^2 + \xi R),$$

where

$$\alpha = \frac{2}{\kappa^2}, \quad \beta = -\frac{\kappa^2}{2w^4}, \quad \gamma = \frac{\kappa^2 \mu}{2w^2}, \quad \zeta = -\frac{\kappa^2 \mu^2}{8};$$

$$\eta = \frac{\kappa^2 \mu^2 (1 - 4\lambda)}{32(1 - 3\lambda)}, \quad \xi = \frac{\kappa^2 \mu^2 \Lambda}{8(1 - 3\lambda)},$$

$$\sigma = -\frac{3\kappa^2 \mu^2 \Lambda^2}{8(1 - 3\lambda)} \tag{5.8}$$

are constant parameters of the theory.

For g_{0i}, we find

$$\frac{\delta S}{\delta g_{0l}} = \frac{\delta S}{\delta N_l} = 2\alpha \nabla_k (K^{kl} - \lambda K g^{kl}) - T^{0l} = 0. \tag{5.9}$$

Finally, for g_{ij} one obtains

$$G_{ij} = T_{ij}, \tag{5.10}$$

where

$$G_{ij} = G_{ij}^{(1)} + G_{ij}^{(2)} + G_{ij}^{(3)} + G_{ij}^{(4)} + G_{ij}^{(5)} + G_{ij}^{(6)}. \tag{5.11}$$

Here, with $\square \equiv \nabla^2$, one has

$$G_{ij}^{(1)} = 2\alpha N K_{ik} K_j^k - \frac{\alpha N}{2} K_{kl} K^{kl} g_{ij} + \alpha (K_{ik} N_j)^{;k} + \alpha (K_{jk} N_i)^{;k}$$
$$\qquad - \alpha (K_{ij} N_k)^{;k} + (i \leftrightarrow j),$$

$$G_{ij}^{(2)} = -2\alpha \lambda N K K_{ij} + \frac{\alpha \lambda N}{2} K^2 g_{ij} - \frac{\alpha \lambda}{\sqrt{g}} g_{ik} g_{jl} \frac{\partial}{\partial t} (\sqrt{g} K g^{kl})$$
$$\qquad - \alpha \lambda (K g_{ik} N_j)^{;k} - \alpha \lambda (K g_{jk} N_i)^{;k} + \alpha \lambda (K g_{ij} N_k)^{;k} + (i \leftrightarrow j),$$

$$G_{ij}^{(3)} = N \xi R_{ij} - \frac{N}{2} (\xi R + \sigma) g_{ij} - \xi N_{;ij} + \xi \square N g_{ij} + (i \leftrightarrow j),$$

$$G_{ij}^{(4)} = 2N \eta R R_{ij} - \frac{N}{2} \eta R^2 g_{ij} + 2\eta \square (N R) g_{ij} - 2\eta (N R)_{;ij} + (i \leftrightarrow j),$$

$$G_{ij}^{(5)} = \square (N(\zeta R_{ij} + \frac{\gamma}{2} C_{ij})) - (N(\zeta R_{ki} + \frac{\gamma}{2} C_{ki}))_{;j}^{\;k}$$
$$\qquad + (N(\zeta R^{kl} + \frac{\gamma}{2} C^{kl}))_{;lk} g_{ij} + (i \leftrightarrow j),$$

and

$$G_{ij}^{(6)} = \frac{1}{2} \frac{\epsilon^{mkl}}{\sqrt{g}} \Big[(Q_{mi})_{;kjl} + (Q_m^{\;n})_{;kin} g_{jl}$$
$$\qquad - (Q_{mi})_{;kn}^{\;\;n} g_{jl} - (Q_{mi})_{;k} R_{jl} - (Q_{mi} R_k^n)_{;n} g_{jl}$$
$$\qquad + (Q_{\;m}^n R_{ki})_{;n} g_{jl} + \frac{1}{2} (R_{\;pkl}^n Q_m^{\;p})_{;n} g_{ij} + Q_{mi} R_{jl;k} \Big]$$
$$\qquad + 2N \zeta R_{ik} R_j^k$$
$$\qquad - \frac{N}{2} (\beta C_{kl} C^{kl} + \gamma R_{kl} C^{kl} + \zeta R_{kl} R^{kl}) g_{ij} - \frac{1}{2} Q_{kl} C^{kl} g_{ij}$$
$$\qquad + (i \leftrightarrow j). \tag{5.12}$$

These equations are very involved. However, already now we can indicate some situations where the equations are essentially simplified.

First of all, it is clear that the equations of motion are simplified if a variable to be found depends only on one argument (examples of such variables are the scale factor for the FRW metric and the radial function for the static spherically symmetric metric). Second, the case of a diagonal metric simplifies the system immediately since one has $N_i = 0$ and $N = \sqrt{|g_{00}|}$. We note that the static spherically symmetric metric (for example, a non-rotating black hole) and FRW metric are diagonal. As for the Gödel-type metric, it has been considered in a tetrad base which strongly simplifies calculations (see [145] for details). Now, let us consider examples of solutions for gravitational field equations.

5.3 Exact Solutions

Let us consider the exact solutions. Again, as earlier, we consider three examples—cosmological FRW metric, black hole and Gödel-type metric.

We follow [144]. So, for the cosmological case, one suggests $N = N(t)$, $N_i = 0$ (since the FRW metric is diagonal) and $g_{ij} = a^2(t)\gamma_{ij}$, where γ_{ij} is the maximally symmetric spatial metric yielding constant scalar curvature: $R = 6k$, and $R_{ij} = 2k\gamma_{ij}$, therefore $\nabla_i R = 0$, and the Cotton tensor is also zero, $C_{ij} = 0$. The matter is suggested to be the function of time only, $\Phi = \Phi(t)$. We can introduce the new Hubble parameter $H = \frac{\dot{a}}{Na}$, where $a = a(t)$ is the usual scale factor in (1.6).

It is natural to suggest that the matter is given by a scalar field which, as usual in cosmology, depends only on time. As a result, the equation of motion for N looks like

$$3\alpha(3\lambda - 1)H^2 + \sigma + \frac{6k\xi}{a^2} + \frac{12k^2(\zeta + 3\eta)}{a^4} = \frac{\dot{\Phi}^2}{N^2} + V(\Phi). \qquad (5.13)$$

For g_{ij}, one finds

$$2\alpha(3\lambda - 1)(\dot{H} + \frac{3}{2}H^2) + \sigma + \frac{2k\xi}{a^2} - \frac{4k^2(\zeta + 3\eta)}{a^4} = -\frac{\dot{\Phi}^2}{N^2} + V(\Phi). \qquad (5.14)$$

Finally, for a matter the equation is

$$\frac{1}{N}\partial_t\left(\frac{\dot{\Phi}}{N}\right) + 3H\frac{\dot{\Phi}}{N} + \frac{1}{2}V_\Phi = 0. \qquad (5.15)$$

One can verify that cyclic or bouncing solutions are possible [146]. In the vacuum case, one can prove directly the possibility of static solutions, while in the presence of the matter, the solutions can be obtained only numerically [147].

We can find static spherically symmetric solutions described by Eq. (3.20). Clearly, the possibility of black holes is of special interest. We start with the particular case of the metric (3.20):

$$ds^2 = -f(r)dt^2 + \frac{dr^2}{f(r)} + r^2(d\theta^2 + \sin^2\theta d\phi^2). \qquad (5.16)$$

It is clear that the Schwarzschild and Reissner-Nordstrom metrics match this form. In [148] it has been explicitly shown that, for $\lambda = 1$, one has

$$f(r) = 1 + \omega r^2 - \sqrt{r(\omega^2 r^3 + 4\omega M)}. \tag{5.17}$$

Here, the ω is a function of constant parameters of the theory. The essential conclusion is that at large distances, i.e. $r \gg (M/\omega)^{1/3}$, one has $f(r) \simeq 1 - \frac{2M}{r} + O(r^{-4})$, that is, the asymptotically Schwarzschild result, i.e. the consistency with the general relativity is achieved.

It has been demonstrated in [148] that the equation $f(r) = 0$ has two solutions, so this black hole has two horizons with $r_{pm} = M(1 \pm \sqrt{1 - \frac{1}{2\omega M^2}})$. The naked singularity is avoided at $\omega M^2 \geq 1/2$.

Now, let us consider the Gödel-type solution (1.10). It has been studied in detail in [145]. First of all, we note that $g_{\phi\phi} = D^2 - H^2 = G(r)$ (other two components of g_{ij} are 1), and $N = \frac{D(r)}{\sqrt{G(r)}}$. So, the positiveness of $G(r)$, and hence satisfying the causality condition, is necessary to have a consistent (real) value of N!

After some change of variables discussed in [145], we can rewrite this metric as

$$ds^2 = -(dt' + \frac{2\Omega}{m} e^{mx} dy)^2 + e^{2mx} dy^2 + dr^2 + dz'^2, \tag{5.18}$$

with $G(x) = v^2 e^{2mx} > 0$, and $v^2 = 1 - \frac{4\Omega^2}{m^2}$, so, the causality is guaranteed if $v^2 > 0$. For this metric, $R_{1212} = -m^2 v^2 e^{2mx}$, $K_{12} = -v\Omega e^{mx}$, $C^{ij} = 0$ and $R = -2m^2$.

To verify the consistency of this solution, we choose the fluid-like matter with

$$T^{\mu\nu} = (p + \rho)u^\mu u^\nu + pg^{\mu\nu}. \tag{5.19}$$

Namely, this matter has been used in the original paper [4]. After solving the algebraic equations, we find $m^2 = \frac{2}{3}\Omega^2$ or $m^2 = \frac{1}{4}\Omega^2$. However, both these solutions appear to be not completely satisfactory since they are non-causal and, moreover, imply in the imaginary value of v (we note again that, as has been proved in [5], the causality is achieved for $m^2 \geq 4\Omega^2$). As for constant parameters of the theory, namely λ, μ and Λ, they can also be obtained in terms of m, Ω, p and ρ; the explicit values are given in [144].

Therefore, we have seen that these solutions of GR are consistent within the HL gravity, at least asymptotically. Again, it is important to note that the HL gravity is power-counting renormalizable (although up to now there are no examples of full-fledged quantum calculations in the theory). Nevertheless, it must be noted that it also displays some difficulties which we will discuss now.

5.4 Modified Versions of HL Gravity

While the HL gravity seems to solve the problem of renormalizability, and the most important classical solutions in it reproduce those for the GR in certain limits, the consistent description of degrees of freedom in HL gravity turns out to be problematic. This fact has been firstly described in [149]. Following that paper, the main problem of the HL gravity looks as follows: the full-fledged general covariance group is broken up to its subgroup which leaves the space-time foliation to be invariant. In other words, since there is no more symmetry between space and time, one has the reduced gauge group for transformations of spatial coordinates only. Thus, the gauge symmetry is partially broken, which implies arising of new degrees of freedom which can imply unstable vacuum, strong coupling and other unusual effects [150]. It was claimed in [149] that, in fact, the extra mode appears to satisfy the first-order equation of motion and hence does not propagate.

To illustrate this fact, let us consider the equations of motion (5.6), (5.9) and (5.10). As we already noted, they are invariant under three-dimensional gauge transformations in the linearized case looking like $\delta g_{ij} = \partial_i \xi_j + \partial_j \xi_i$. These transformations allow to impose the gauge $N_i = 0$ [149, 150]. Afterwards, Eq. (5.3) takes the form: $\dot{g}_{ij} = 2N K_{ij}$. However, in the system (5.6), (5.9) and (5.10) there is no equation for the evolution of N! And since N is separated from all other dynamical variables, it cannot be fixed by gauge transformations. As a result, one concludes that N describes the new degree of freedom. To study it, we take the time derivative of (5.7), combine it with other equations and arrive at

$$\nabla_i \left(N^2 \left[\xi (\lambda - 1) \nabla^i K + F^i (K_{jk}, R_{jk}, K) \right] \right) = 0. \tag{5.20}$$

It is easy to see that we have 13 dynamical variables (K_{ij}, g_{ij}, N) and five constraints given by (5.6), (5.9) and (5.20), so, we rest with 8 independent variables. Using three gauge parameters ξ_i, we can eliminate three more variables. For five remaining ones, we have four initial conditions for two helicities of h_{ij}. So, we stay with one extra degree of freedom!

More detailed analysis performed in [149] shows that if we consider k_{ij}, a small fluctuation of K_{ij}, its trace $\kappa = k_i^i$ does not propagate since $\nabla^2 \kappa = 0$. So, we can conclude that this extra mode is non-physical.

Returning to the dynamics of N, we can fix N through the additive term in the action given by $S_n = \int d^3x dt \sqrt{g} N \frac{\rho}{2} (N^{-2} - 1)$, which implies strong coupling (roughly speaking, due to the presence of the constraint). Under some tricks like covariant extension (i.e. introducing a Lorentz-covariant analogue), it appears to be similar to the Einstein-aether action (with ϕ being a Stuckelberg field) $S_n = \int d^3x dt \sqrt{g} \frac{\rho}{2} (\nabla_\mu \phi \nabla^\mu \phi - 1)$ [151]; ϕ is called chronon since there is a gauge in which this field is equal to a time coordinate, $\phi = t$.

It was argued in [151] that if we introduce $u_\mu = \frac{\partial_\mu \phi}{\sqrt{X}}$, with $X = g_{\mu\nu} \partial^\mu \phi \partial^\nu \phi$, we can add some terms to our action to get a consistent theory! Actually, we have

$$S = -\frac{1}{\kappa^2} \int d^4x \sqrt{|g|} (R_4 + (\lambda - 1)(\nabla_\mu u^\mu)^2$$
$$+ \alpha u^\mu (\nabla_\mu u^\nu) u^\lambda (\nabla_\lambda u_\nu) + \ldots), \tag{5.21}$$

and this action, for splitting $\phi \to t + \chi$, yields reasonable dispersion relations for χ like $\omega^2 = C\vec{p}^2$, with C being some number. In [151], also some cosmological impacts of this term were studied. An aside result is an emergence of Einstein-aether action. So, the consistent extension of the HL gravity is found.

Another approach is based on the use of so-called projectable version of the HL gravity, where the lapse N is suggested to be a function of time only, $N = N(t)$. However, it turns out that although in this case the theory is strongly simplified, the scalar excitation is still unstable and cannot be ruled out [152].

5.5 Conclusions

Let us make some conclusions regarding the HL gravity. As we already noted, the key idea of the HL gravity is that the usual general covariance is an essentially low-energy phenomenon but not a fundamental feature of the nature. In a certain sense, it can be said that the HL concept was developed to "sacrifice" general covariance in order to conciliate desired renormalizability with absence of ghosts. In this context, it should be noted that the breaking of general covariance in gravity is discussed as well in "usual" LV gravity models without strong space-time asymmetry [111].

We demonstrated how the known GR solutions are modified within the HL context. Within the cosmological context, accelerated and bouncing solutions are possible; thus the HL gravity is a good candidate to solve the dark energy problem. We demonstrated that there are black hole solutions behaving like usual Schwarzschild BHs at large distances. Also, we demonstrated that the Gödel-type solutions consistent within the HL gravity are non-causal, but one should note that the Gödel solution itself is non-causal.

However, a quantum-mechanical description of the HL gravity is rather problematic. One of the reasons is the very complicated structure of the classical action potentially implying a very large number of divergent contributions; therefore while the HL gravity is power-counting renormalizable, detailed study of its renormalization is still to be done multiplicatively renormalizable. Another difficulty is the question about an extra degree of freedom. While it was in principle solved in [151], where the "healthy extension" of HL gravity was introduced, the problem now consists in obtaining physically consistent results on the base of this extension. Therefore, even in this case we have more questions than answers. We close the discussion recommending an excellent review of HL gravity presented in [153].

Chapter 6
Nonlocal Gravity

6.1 Motivations

As we have noted several times along this review, the main problem of various gravity models is the development of their consistent quantum description. Indeed, the Einstein gravity is non-renormalizable, and the introduction of higher derivative additive terms implies the arising of ghosts. We have argued in the previous chapter that the HL gravity seems to be a good solution since it is power-counting renormalizable, and ghosts are absent since the action involves only second time derivatives. However, the HL gravity, first, is very complicated, and, second, breaks the Lorentz symmetry strongly, and, third, displays a problem of extra degrees of freedom whose solving, as we noted, requires special efforts. At the same time, the concept of nonlocality developed originally within a phenomenological context in order to describe finite-size effects (see, for example, [154]) began to attract interest. Besides this, the nonlocality enjoys also a stringy motivation since the factors like $e^{\alpha \Box}$ emerge naturally within the string context [155]. The key idea of nonlocal field theories looks as follows. Let us consider for example the free scalar field whose Lagrangian is

$$\mathcal{L} = \frac{1}{2}\phi f(\Box/\Lambda^2)\phi, \tag{6.1}$$

where $f(z)$ is a non-polynomial function called the form factor (with Λ being the characteristic nonlocality scale) which we choose to satisfy the following requirements.

First, at small arguments this function should behave as $f(z) = a + z$, in order to provide the correct IR asymptotic behavior $\propto \Box + m^2$. Second, the $(f(z))^{-1}$, describing the propagator of the theory, must decay rapidly at $|z| \to \infty$ (in principle, we can consider only Euclidean space, so, z can be treated as an essentially positive variable), so that integrals like $\int_0^\infty (f(z))^{-1} z^n dz$ must be finite at any non-negative n, to guarantee finiteness of the theory (in principle in some cases this requirement is weakened, and the corresponding theory is required to be not finite but only (super)renormalizable, as it occurs in the theory studied in [156]). Third, the $f(z)$ is required to be so-called entire function, i.e. it cannot be presented in the form of a product of primitive multipliers like $(z - a_1)(z - a_2)\ldots$, so, its propagator has no different poles (as we noted in Chap. 2, namely the presence of such a set of poles implies the existence of ghost modes). The simplest example of such a function is the exponential, $f(z) = e^z$ (see [157, 158]). However, other entire functions can be used as well, for example, e^{z^2} and more involved ones. An example of a form factor function consistent with the super-renormalizability condition, i.e. an asymptotically polynomial one, although of a rather involved form, is presented in [156]; we note that such a function possesses some advantages namely within the gravity context where, unlike scalar or gauge nonlocal theories, the exponential form factor would imply the presence of exponentially growing momentum depending functions in vertices making the situation with renormalization of the theory much more complicated (see [159] and references therein for a general discussion of nonlocal form factors). In principle, in many interesting cases the explicit expression for the form factor does not require to be specified.

Other motivations for nonlocality are the loop quantum gravity dealing with finite-size objects, and the noncommutativity, where the Moyal product is essentially non-local by construction. At the same time, it is interesting to note that although the so-called coherent states approach [160] has been motivated by quantum mechanics, by its essence it represents itself as a natural manner to implement nonlocality, so that all propagators carry the factor $e^{-\theta k^2}$, with θ being the nonlocality scale. Within the gravity context, use of the nonlocal methodology appears to be especially promising since it is expected that the nonlocality, being implemented in a proper manner, can allow to achieve renormalizability without paying the price of arising the ghosts. The first step in this study has been done in the seminal paper [157].

6.2 Some Results in Non-gravitational Nonlocal Theories

Before embarking on studies of nonlocal gravity, let us first discuss the most interesting results in non-gravitational nonlocal theories, especially within the context of quantum corrections.

As we already noted, effectively the nonlocal methodology has been applied to perturbative studies for the first time within the coherent states approach [160] which includes Gaussian propagator guaranteeing the convergence of quantum corrections, in non-gravitational theories. Further, various other studies have been performed. An important role was played by the paper [161] where the effective potential in a nonlocal theory has been studied for the first time. In that paper, the following theory has been introduced:

$$\mathcal{L} = -\frac{1}{2}\phi(\exp(\Box/\Lambda^2)\Box + m^2)\phi - V(\phi). \tag{6.2}$$

Here, Λ is a characteristic nonlocality scale. For this theory, one can calculate the one-loop effective potential given by the following integral:

$$V^{(1)} = \frac{i}{2} \int \frac{d^4k}{(2\pi)^4} \ln\left(-\exp\left(-\frac{k^2}{\Lambda^2}\right)k^2 + m^2 + V''\right). \qquad (6.3)$$

It is clear that in the limit $\Lambda^2 \to \infty$, the theory is reduced to a usual one. The exponential factors guarantee finiteness. It is easy to see that there are no ghosts in the theory since there is no product of different denominators $\Box + m_i^2$ in the propagator of the theory (in other words, the function $f(k^2) = -\exp(-\frac{k^2}{\Lambda^2})k^2 + m^2$, characterizing the denominator of the free propagator, is entire). However, the integral (6.3) can be calculated only approximately for various limits (namely, the V'' is considered to be either very large or very small), and it is easy to see that it is singular at $\Lambda \to \infty$. In [161], a procedure to isolate this singularity, based on the normal-ordering prescription, has been adopted, and some estimations for the one-loop effective potential were done; although it was not calculated explicitly, namely the resulting nonlocal contribution to the effective potential was shown to be small in comparison with the effective potential calculated in a corresponding local field theory model. Further, the calculation of the one-loop effective potential for the superfield theories representing themselves as various nonlocal extensions of the Wess-Zumino model and super-QED, with the form factors being the exponential functions, was performed in [162, 163], where, unlike [161], no normal-ordering prescription was applied. The one-loop Kählerian effective potential, by definition depending only on chiral matter superfields but not on their derivatives (for a description of the superfield formalism and related definitions and conventions, see, for example, [15]), has been obtained explicitly for these models. For example, for the theory, representing itself as a nonlocal generalization of the Wess-Zumino model and given by the superfield action

$$S = \int d^8z\, \bar{\Phi} e^{-\Box/\Lambda^2}\Phi + \left[\int d^6z \left(\frac{m}{2}\Phi^2 + \frac{\lambda}{3!}\Phi^3\right) + h.c.\right], \qquad (6.4)$$

where Φ is a chiral superfield and $\bar{\Phi}$ is an antichiral one, $d^8z = d^4x\, d^2\theta\, d^2\bar{\theta}$ is the full superspace measure and $d^6z = d^4x\, d^2\theta$ is the chiral superspace measure, the one-loop Kählerian effective potential looks like

$$K^{(1)} = -\frac{\bar{\Psi}\Psi}{32\pi^2} \ln \frac{2\bar{\Psi}\Psi}{e^{1-\gamma}\Lambda^2} + O\left(\frac{1}{\Lambda^2}\right), \qquad (6.5)$$

where $\Psi = m + \lambda\Phi$ and $\bar{\Psi} = m + \lambda\bar{\Phi}$. We see that the result is singular at $\Lambda \to \infty$. Indeed, if in this model (and other nonlocal models as well), one considers the limit of an infinite nonlocality scale $\Lambda \to \infty$, the theory returns to the local limit and becomes divergent, i.e. the nonlocality acts as a kind of the higher derivative regularization, so, the quantum contributions are singular in this limit growing as Λ^2 if the local counterpart of the theory involves quadratic divergences, or as $\ln \Lambda^2$, if it involves the logarithmic ones. From a formal viewpoint, the existence of these singularities can be exemplified by the fact that the typical integral in nonlocal (Euclidean) theory grows quadratically with Λ scale since $\int \frac{d^4k}{(2\pi)^4} \frac{1}{k^2} e^{-k^2/\Lambda^2} \propto \Lambda^2$. Effectively, the problem of the singularity of the result at $\Lambda \to \infty$ is nothing more that the problem of large quantum corrections arising also in higher derivative and noncommutative field theories.

At the same time, the problems of unitarity and causality in nonlocal theories require special attention since the nonlocality is commonly associated with an instant propagation of a signal. These problems were discussed in detail in various papers. So, it has been claimed in [156] that the problems of unitarity and causality can be solved at least for certain forms of nonlocal functions. Afterwards, this result was corroborated and discussed in more detail in [164]. Further discussions of the problem of the unitarity in nonlocal field theory models are presented in [165] and [166], where Euclidean and Minkowskian prescriptions, respectively, were used for calculating the Feynman diagrams, and it has been argued that while the Euclidean formulation is consistent with the unitarity requirement, the Minkowskian one appears to be non-unitary. However, the complete discussion of unitarity and causality in nonlocal field theories is still to be done. Alternatively, the nonlocal theories must be treated only as effective ones.

So, to go to studies of gravity, we can formulate some preliminary conclusions: (i) there is a mechanism allowing to avoid UV divergences; (ii) this mechanism is Lorentz covariant and ghost-free; (iii) the unitarity and causality still are to be studied.

6.3   Classical Solutions in Nonlocal Gravity Models

So, let us introduce examples of nonlocal gravity models. The paradigmatic example has been proposed in [158], where the Lagrangian $\mathcal{L} = \frac{1}{2\kappa^2}\sqrt{|g|}F(R)$ was studied, with

$$F(R) = R - \frac{R}{6}\left(\frac{e^{-\Box/M^2} - 1}{\Box}\right)R. \tag{6.6}$$

Here, the d'Alembertian operator $\Box$ is covariant: $\Box = g^{\mu\nu}\nabla_\mu\nabla_\nu$. This is the nonlocal extension of R^2-gravity.

First of all, it is easy to show that this theory is ghost-free. Indeed, we can expand

$$F(R) = R + \sum_{n=0}^{\infty} \frac{c_n}{M^{2n+2}} R\Box^n R, \tag{6.7}$$

with $c_n = -\frac{1}{6}\frac{(-1)^{n+1}}{(n+1)!}$. We can rewrite this Lagrangian with auxiliary field Φ and scalar ψ (we can eliminate first Φ, and then ψ, through their equations of motion):

$$\mathcal{L} = \frac{1}{2\kappa^2}\sqrt{|g|}\left(\Phi R + \psi \sum_{n=1}^{\infty}\frac{c_n}{M^{2n+2}}\Box^n\psi - \left[\psi(\Phi - 1) - \frac{c_0}{M^2}\psi^2\right]\right). \quad (6.8)$$

Then we do conformal transformations $g_{mn} \rightarrow \Phi g_{mn}$, with $\Phi \simeq 1 + \phi$, to absorb Φ in curvature term. As a result, we arrive at the Lagrangian

$$\mathcal{L} = \frac{1}{2\kappa^2}\sqrt{|g|}\left(R + \psi \sum_{n=0}^{\infty}\frac{c_n}{M^{2n+2}}\Box^n\psi - \psi\phi + \frac{3}{2}\phi\Box\phi\right), \quad (6.9)$$

with the equations of motion being

$$\psi = 3\Box\phi; \quad \phi = 2\sum_{n=0}^{\infty}\frac{c_n}{M^{2n+2}}\Box^n\psi. \quad (6.10)$$

From here, we have the equation of motion for ϕ:

$$\left(1 - 6\sum_{n=0}^{\infty}c_n\frac{\Box^{n+1}}{M^{2n+2}}\right)\phi = e^{-\Box/M^2}\phi = 0. \quad (6.11)$$

The l.h.s. is evidently entire, so we have no ghosts.

We conclude that the nonlocality in the geometric sector can be transferred to the matter sector! This is valid for various models. In a certain sense, this fact is analogous to the observation made in Sect. 2.3 where it was argued that the $f(R)$ gravity, representing itself as an example of higher derivative theory, can be mapped to a scalar-tensor gravity with no higher derivatives in the geometric sector.

The Lagrangian (6.6) can be rewritten as [167]

$$\mathcal{L} = \sqrt{|g|}\left(\frac{1}{2\kappa^2}R + \frac{\lambda}{2}RF(\Box)R - \Lambda + \mathcal{L}_M\right). \quad (6.12)$$

The function $F(\Box)$ is assumed to be analytic, as it is motivated by string theory, and, moreover, in the analytic case the theory does not display problems in IR limit. The Gaussian case, which is especially convenient from the viewpoint of the UV finiteness, is the perfect example. The equations of motion take the form

$$\left[\frac{1}{2\kappa^2} + 2\lambda F(\Box)R\right] G_\nu^\mu = T_\nu^\mu + \Lambda\delta_\nu^\mu + \lambda K_\nu^\mu - \frac{\lambda}{2}(K_\alpha^\alpha + K_1) -$$

$$-\frac{\lambda}{2} RF(\Box)R\delta_\nu^\mu + 2\lambda(g^{\mu\alpha}\nabla_\alpha\nabla_\nu - \delta_\nu^\mu\Box)F(\Box)R, \qquad (6.13)$$

$$K_\nu^\mu = g^{\mu\rho}\sum_{n=1}^{\infty} f_n \sum_{l=0}^{n-1} \partial_\rho\Box^l R\,\partial_\mu\Box^{n-l-1}R;$$

$$K_1 = \sum_{n=1}^{\infty} f_n \sum_{l=0}^{n-1} \Box^l R\,\Box^{n-l}R; \qquad F(\Box) = \sum_{n=0}^{\infty} f_n\Box^n.$$

It is important to note that in the two last lines $\Box^l$ acts *only* to the adjacent R.

Now, the natural problem is to find some solutions of these equations. In [167], the following ansatz has been proposed, with r_1, r_2 being some real numbers:

$$\Box R - r_1 R - r_2 = 0 \qquad (6.14)$$

which implies (here f_0 is of zeroth order in the expansion of $F(\Box)$ in series)

$$F(\Box)R = F(r_1)R + \frac{r_2}{r_1}(F(r_1) - f_0). \qquad (6.15)$$

This allows to reduce the order of equations to at maximum second. It is clear that constant curvature makes the equation trivial; just this situation occurs for Gödel-type solutions.

One can find nontrivial cosmological solutions for this theory. In particular, bouncing solutions, for $r_1 > 0$, are possible:

$$a(t) = a_0 \cosh\left(\sqrt{\frac{r_1}{2}}t\right). \qquad (6.16)$$

Let us discuss cosmological solutions in this theory in more detail. Indeed, if we substitute the FRW metric (1.6) to (6.13), and suggest that, as usual in cosmology, $\rho = \rho_0(\frac{a_0}{a})^4$, we have from (6.14), with $r_1 \neq 0$:

$$\frac{d^3 H}{dt^3} + 7H\ddot{H} + 4\dot{H}^2 - 12H^2\dot{H} = -2r_1 H^2 - r_1\dot{H} - \frac{r_2}{6}, \qquad (6.17)$$

whose solution is $H = \sqrt{\frac{r_1}{2}}\tanh\left(\sqrt{\frac{r_1}{2}}t\right)$ which just implies the hyperbolic dependence of $a(t)$ (6.16). It is well known that namely such a scenario (decreasing of scale factor changing then to increasing) is called the bouncing one. We also introduce $h_1 = \ddot{H}/M^3$ (we note that this object does not differ essentially from a constant being a second derivative from a slowly varying function).

The density can be found as well: if we redefine $F(\Box) \to F(\Box/M^2)$, with M being the characteristic nonlocality scale, we find

$$\rho_0 = \frac{3(\frac{1}{\kappa^2}r_1 - 2\lambda f_0 r_2)(r_2 - 12h_1 M^4)}{12r_1^2 - 4r_2}.$$ (6.18)

Let us discuss possible implications of the equation (6.17). The cosmological constant turns out to be equal to $\Lambda = -\frac{r_2}{4\kappa^2 r_1}$, and there are three scenarios for the evolution of the Universe:

1. $\Lambda < 0, r_1 > 0, r_2 > 0$—cyclic Universe (in particular one can have cyclic inflation).
2. $\Lambda > 0, r_1 < 0, r_2 > 0$—first contraction, then very rapid inflation (superinflation) $a(t) \propto \exp(kt^2)$.
3. $\Lambda > 0, r_1 > 0, r_2 < 0$—constant curvature $R = 4\frac{\Lambda}{M_P^2}$, i.e. de Sitter solution.

So we find that accelerating solutions are possible within all these scenarios. Again, we note that in the constant scalar curvature case, we have drastic reducing of equations.

Moreover, it has been shown in [168] that for $\mathcal{L} = \sqrt{|g|}\sqrt{R - 2\Lambda} F(\Box)\sqrt{R - 2\Lambda}$, with $F(\Box)$ being an arbitrary analytic function, there are hyper-exponentially accelerating cosmological solutions $a(t) \propto e^{kt^2}$.

The next step in the study of nonlocal theories consists in introducing non-analytic functions of the d'Alembertian operator. The simplest case is $F(\Box) = \frac{1}{\Box}$. Actually it means that we must consider terms like $R\Box^{-1}R$. It is clear that the gravity extension with such a term is non-renormalizable since the propagator behaves as only $\frac{1}{k^2}$, so we gain nothing in comparison with the usual Einstein-Hilbert gravity [169]. However, theories with negative degrees of the d'Alembertian operator can display new tree-level effects, especially within the cosmological context where an important class of nonlocal gravity models has been introduced in [170]. The action of this class of theories is

$$S = \int d^4 x \sqrt{|g|} \left(\frac{1}{2\kappa^2} \left(R + Rf(\Box^{-1}R) - 2\Lambda \right) + \mathcal{L}_m \right).$$ (6.19)

We note that the presence of the factor $\Box^{-1}$ actually implies "retarded" solutions behaving similar to the potential of a moving charge in electrodynamics. Further, this action has been considered in [171], and below, we review the discussion given in that paper.

It is convenient to rewrite the action (6.19) with the use of two extra scalar fields ξ and η:

$$S = \int d^4 x \sqrt{|g|} \left[\frac{1}{2\kappa^2} [R(1 + f(\eta) - \xi) + \xi\Box\eta - 2\Lambda] + \mathcal{L}_m \right].$$ (6.20)

Varying this action with respect to ξ and expressing $\eta = \Box^{-1}R$, we return to (6.19). This corroborates the already mentioned idea that the modified gravity is in many cases equivalent to a scalar-tensor gravity.

Then, we vary (6.20) with respect to the metric and η respectively:

$$\Box \xi + f_\eta(\eta) R = 0;$$

$$\frac{1}{2} g_{\mu\nu}[R(1 + f(\eta) - \xi) - \partial_\alpha \xi \partial^\alpha \eta - 2\Lambda] - R_{\mu\nu}(1 + f(\eta) - \xi) +$$

$$+ \frac{1}{2}(\partial_\mu \xi \partial_\nu \eta + \partial_\mu \eta \partial_\nu \xi) - (g_{\mu\nu}\Box - \nabla_\mu \nabla_\nu)(f(\eta) - \xi) = -G T_{\mu\nu}. \quad (6.21)$$

We consider the FRW cosmological metric (1.6) with $k = 0$. As usual, the Hubble parameter is $H = \frac{\dot{a}}{a}$. The evolution equation for matter is usual:

$$\dot{\rho} = -3H(\rho + p). \quad (6.22)$$

For scale factor and scalars, we have

$$2\dot{H}(1 + f(\eta) - \xi) + \dot{\xi}\dot{\eta} + \left(\frac{d^2}{dt^2} - H\frac{d}{dt}\right)(f(\eta) - \xi) + G(\rho + p) = 0;$$

$$\ddot{\eta} + 3H\dot{\eta} = -6(\dot{H} + 2H^2);$$

$$\ddot{\xi} + 3H\dot{\xi} = -6(\dot{H} + 2H^2) f_\eta(\eta). \quad (6.23)$$

We start with the de Sitter space corresponding to $H = H_0 = const$, with the scalar curvature being $R = 12H_0^2$. The equation of state is $p = \omega\rho$, as usual, so, we have the following solutions for the scalar η and the density:

$$\eta(t) = -4H_0(t - t_0) - \eta_0 e^{-3H_0(t-t_0)};$$

$$\rho(t) = \rho_0 e^{3(1+\omega)H_0 t}. \quad (6.24)$$

Then we introduce the new variable $\Psi = f(\eta) - \xi$, and its equation of evolution is

$$\ddot{\Psi} + 5H_0\dot{\Psi} + 6H_0^2(1 + \Psi) - 2\Lambda + G(\omega - 1)\rho = 0. \quad (6.25)$$

For η we have

$$\dot{\eta}^2 f_{\eta\eta} + (\ddot{\eta} + 3H_0\dot{\eta} - 12H_0^2)f_\eta = \ddot{\Psi} + 3H_0\dot{\Psi}. \quad (6.26)$$

This equation is a necessary condition for the existence of the de Sitter solution.

Let us consider the particular case $\eta_0 = 0$ in (6.24). So, (6.26) reduces to

$$16H_0^2 f_{\eta\eta} - 24H_0^2 f_\eta = \ddot{\Psi} + 3H_0\dot{\Psi}. \quad (6.27)$$

So, knowing Ψ, one can find $f(\eta)$. It remains to solve (6.25). Some characteristic cases are

- $\rho_0 = 0$: $\Psi = C_1 e^{-3H_0 t} + C_2 e^{-2H_0 t} - 1 + \frac{\Lambda}{3H_0^2}$;
- $w = 0$: $\Psi = C_1 e^{-3H_0 t} + C_2 e^{-2H_0 t} - 1 + \frac{\Lambda}{3H_0^2} - \frac{G\rho_0}{H_0} e^{-3H_0 t} t$.

- $w = -1/3$: $\Psi = C_1 e^{-3H_0 t} + C_2 e^{-2H_0 t} - 1 + \frac{\Lambda}{3H_0^2} + \frac{4G\rho_0}{3H_0} e^{-2H_0 t} t$.

As for the function $f(\eta)$, in all cases it will be proportional to $e^{\eta/\beta}$, with $\beta > 0$ (or, at most, linear combination of such functions with various values of β). Effectively, we demonstrated arising of the exponential potential widely used in cosmology.

An important particular case is $\eta_0 = 0$. It follows from (6.24) that we have for $\beta \neq 4/3$:

$$\xi = -\frac{3 f_0 \beta}{3\beta - 4} e^{-H_0(t-t_0)/\beta} + \frac{c_0}{3H_0} e^{-3H_0(t-t_0)} - \xi_0;$$

$$\eta = -4H_0(t - t_0); \quad \omega = \frac{4}{3\beta} - 1, \quad \Lambda = 3H_0^2(1 + \xi_0);$$

$$\rho_0 = \frac{6(\beta - 2) H_0^2 f_0}{\beta \kappa^2}, \tag{6.28}$$

so we can have exotic matter for $0 < \beta < 2$. And at $\beta = 2$ we have vacuum. If $\beta = 4/3$, we have $\omega = 0$, and $\rho < 0$ (ghost-like dust).

However, we note that the nonlocal modifications of gravity are used mostly in cosmology. One of a few discussions of other metrics within the nonlocal gravity has been presented in [172] where not only cosmological but also (anti) de Sitter-like and some of black hole solutions (Schwarzschild and Kerr ones) and the Gödel metric, and Ricci-flat metrics, in general, were discussed for theories involving, besides the already mentioned term $RF(\Box)R$, also the terms $R_{\mu\nu} F_1(\Box) R^{\mu\nu}$ and $R_{\mu\nu\alpha\beta} F_2(\Box) R^{\mu\nu\alpha\beta}$, with F, F_1, F_2 being some functions of the covariant d'Alembertian operator. However, for the general Gödel-type metric (1.10), the situation is more interesting. In [173], this metric has been considered within the more generic nonlocal gravity involving three terms mentioned in [172]. We note that the Gödel-type metric is not Ricci-flat, therefore, the situation in this case is highly nontrivial.

In this case, we start with the classical action

$$S = \frac{1}{2\kappa^2} \int d^4 x \sqrt{|g|} \left[R - 2\Lambda + \frac{1}{M_N^2} \left(R \mathcal{F}_1(\Box) R + R_{\mu\nu} \mathcal{F}_2(\Box) R^{\mu\nu} + \right. \right.$$
$$\left. \left. + C_{\mu\nu\alpha\beta} \mathcal{F}_3(\Box) C^{\mu\nu\alpha\beta} \right) \right] + S_m[g_{\mu\nu}, \Psi], \tag{6.29}$$

where $\mathcal{F}_{1,2,3}(\Box)$ are three analytic functions of the d'Alembertian operator, $C_{\mu\nu\lambda\rho}$ is the Weyl tensor and Ψ are matter fields.

The equations of motion look like (cf. [174])

$$E^{\alpha\beta} = G^{\alpha\beta} + \Lambda g^{\alpha\beta} + P_1^{\alpha\beta} + P_2^{\alpha\beta} + P_3^{\alpha\beta} - 2\Omega_1^{\alpha\beta} + g^{\alpha\beta}(g_{\mu\nu}\Omega_1^{\mu\nu} + \bar{\Omega}_1) - 2\Omega_2^{\alpha\beta} +$$
$$+ g^{\alpha\beta}(g_{\mu\nu}\Omega_2^{\mu\nu} + \bar{\Omega}_2) - 4\Delta_2^{\alpha\beta} - 2\Omega_3^{\alpha\beta} + g^{\alpha\beta}(g_{\mu\nu}\Omega_3^{\mu\nu} + \bar{\Omega}_3) - 8\Delta_3^{\alpha\beta} =$$
$$= \kappa^2 T^{\alpha\beta}, \tag{6.30}$$

where $G^{\alpha\beta} = R^{\alpha\beta} - \frac{1}{2}Rg^{\alpha\beta}$ is the usual Einstein tensor, the notation $A^{(l)} \equiv \Box^l A$ is used and

$$P_1^{\alpha\beta} = \frac{1}{2M_N^2}\left[(4G^{\alpha\beta} + g^{\alpha\beta}R - 4(\nabla^\alpha\nabla^\beta - g^{\alpha\beta}\Box))\,\mathcal{F}_1(\Box)R\right];$$

$$P_2^{\alpha\beta} = \frac{1}{2M_N^2}\left[4R_\nu^\alpha\,\mathcal{F}_2(\Box)R^{\nu\beta} - g^{\alpha\beta}R^{\mu\nu}\,\mathcal{F}_2(\Box)R_{\mu\nu} - 4\nabla_\nu\nabla^\beta(\mathcal{F}_2(\Box)R^{\nu\alpha}) + \right.$$

$$\left. + 2\Box(\mathcal{F}_2(\Box)R^{\alpha\beta}) + 2g^{\alpha\beta}\nabla_\mu\nabla_\nu(\mathcal{F}_2(\Box)R^{\mu\nu})\right];$$

$$P_3^{\alpha\beta} = \frac{1}{2M_N^2}\left[-g^{\alpha\beta}C^{\mu\nu\sigma\gamma}\,\mathcal{F}_3(\Box)C_{\mu\nu\sigma\gamma} + 4C_{\mu\nu\sigma}^\alpha\,\mathcal{F}_3(\Box)C^{\beta\mu\nu\sigma} - \right.$$

$$\left. -4(R_{\mu\nu} + 2\nabla_\mu\nabla_\nu)(\mathcal{F}_3(\Box)C^{\beta\mu\nu\alpha})\right];$$

$$\Omega_1^{\alpha\beta} = \frac{1}{2M_N^2}\sum_{n=1}^\infty f_{1_n}\sum_{l=0}^{n-1}\nabla^\alpha R^{(l)}\nabla^\beta R^{(n-l-1)}, \quad \bar{\Omega}_1 = \frac{1}{2M_N^2}\sum_{n=1}^\infty f_{1_n}\sum_{l=0}^{n-1}R^{(l)}R^{(n-l)};$$

$$\Omega_2^{\alpha\beta} = \frac{1}{2M_N^2}\sum_{n=1}^\infty f_{2_n}\sum_{l=0}^{n-1}(\nabla^\alpha R^{\mu\nu(l)})(\nabla^\beta R_{\mu\nu}^{(n-l-1)}), \quad \bar{\Omega}_2 = \frac{1}{2M_N^2}\sum_{n=1}^\infty f_{2_n}\sum_{l=0}^{n-1}R^{\mu\nu(l)}R_{\mu\nu}^{(n-l)};$$

$$\Delta_2^{\alpha\beta} = \frac{1}{4M_N^2}\sum_{n=1}^\infty f_{2_n}\sum_{l=0}^{n-1}\nabla^\nu\left(R_{\sigma\nu}^{(l)}\nabla^{(\alpha}R^{\beta)\sigma(n-l-1)} - (\nabla^{(\alpha}R_{\sigma\nu})R^{\beta)\sigma(n-l-1)}\right);$$

$$\Omega_3^{\alpha\beta} = \frac{1}{2M_N^2}\sum_{n=1}^\infty f_{3_n}\sum_{l=0}^{n-1}(\nabla^\alpha C_{\nu\rho\sigma}^{\mu(l)})(\nabla^\beta C_\mu^{\ \nu\rho\sigma(n-l-1)}),$$

$$\bar{\Omega}_3 = \frac{1}{2M_N^2}\sum_{n=1}^\infty f_{3_n}\sum_{l=0}^{n-1}C_{\nu\rho\sigma}^{\mu(l)}C_\mu^{\ \nu\rho\sigma(n-l-1)};$$

$$\Delta_3^{\alpha\beta} = \frac{1}{4M_N^2}\sum_{n=1}^\infty f_{3_n}\sum_{l=0}^{n-1}\nabla^\nu\left(C_{\nu\sigma\mu}^{\rho(l)}\nabla^{(\alpha}C_\rho^{\ \beta)\sigma\mu(n-l-1)} - (\nabla^{(\alpha}C_{\nu\sigma\mu}^{|\rho(l)|})C_\rho^{\ \beta)\sigma\mu(n-l-1)}\right). \quad (6.31)$$

While the equation of motion (6.30) is extremely cumbersome, it is satisfied by the Gödel-type metric. Indeed, if we apply the condition of the space-time homogeneity (1.11), that is, $\frac{H'}{D} = 2\Omega$, $\frac{D''}{D} = m^2$, after using the Newman-Penrose formalism (see details in [173]), one shows that in this case, the scalar curvature is $R = 2(m^2 - \Omega^2) = const$, therefore, besides $R\Box^n R = 0$ for $n \geq 1$, we have

$$R^{\mu\nu}\Box^n R_{\mu\nu} = (-1)^n 6^{n-1}4\omega^{2n}(4\Omega^2 - m^2), \quad \text{for} \quad n \geq 1;$$

$$C^{\mu\nu\alpha\beta}\Box^n C_{\mu\nu\alpha\beta} = (-1)^n 6^{n-1}8\Omega^{2n}(4\Omega^2 - m^2), \quad \text{for} \quad n \geq 1. \quad (6.32)$$

As a result, at $4\Omega^2 - m^2 = 0$ (that is, the RT solution [5]) or at $\Omega = 0$ (zero vorticity case), all "extra" terms in (6.30) vanish, and the modified Einstein equations reduce to the GR ones; thus, we succeeded to show that at least in these two cases the Gödel-type metric solves nonlocal gravity equations. It is important that both these solutions are completely causal (we remind that non-causal ones correspond to $4\Omega^2 - m^2 > 0$ while the case $4\Omega^2 - m^2 \leq 0$ is causal; cf. [5]). Further continuation of this study has been performed in [175], where a discussion of consistency of generic Gödel-type metrics within more general nonlocal gravity models was performed, and the authors claimed that, apparently, causality-violating Gödel-type solutions are ruled out in renormalizable (and unitary, at the same time) nonlocal gravity models.

Let us say a few words about non-analytic nonlocal extensions of gravity. In [176], the additive term $\mu^2 R \square^{-2} R$ was introduced and shown to be consistent with cosmological observations. However, this theory turns out to be problematic from the causality viewpoint [177]. Also, in [178], the first-order correction in μ^2 to the Schwarzschild solution in a theory with this term has been obtained explicitly.

To close the discussion, it is important to note that the nonlocal gravity can arise as an effective theory as a result of integration over some matter fields. Namely in this manner, the term $R \square^{-1} R$ contributes to the trace anomaly, at least in two dimensions, in [21]. Therefore, the presence of nonlocal terms can be apparently treated as a consequence of some hidden couplings with matter.

6.4 Conclusions

We discussed various nonlocal extensions of gravity. The key property of nonlocal theories is the possibility to achieve UV finiteness for an appropriate choice for nonlocal form factor(s). However, apparently explicit quantum calculations in nonlocal gravity models would be extremely complicated from a technical viewpoint; therefore, up to now, all studies of such theories are completely classical ones. Moreover, most papers on nonlocal gravity models are devoted to cosmological aspects of these theories, and the results demonstrated along this chapter allow to conclude that nonlocal extensions of gravity can be treated as acceptable solutions for the dark energy problem. It is worth mentioning that a very interesting discussion of a non-relativistic limit of nonlocal gravity is presented in the excellent book [179]. A massive extension of nonlocal gravity is also possible [58]. At the same time, nonlocal theories, including gravitational ones, display certain difficulties. The main problem is the one of unitarity and causality still requires special attention.

To conclude this chapter, let us emphasize the main directions for the studies of nonlocal gravity models. First, clearly, it will be very important to check the consistency of different known GR solutions, especially, various black holes (including, for example, non-singular and rotating ones). Second, various nonlocal form factors, not only Gaussian ones, are to be introduced, and their impact must be tested within the gravity context. Third, the study of quantum effects in nonlocal gravity models is of special importance since namely at the perturbative level, the main advantages of these theories such as the expected UV finiteness are crucial. It is natural to hope that these studies will be performed in the next years.

Chapter 7
Comments on Non-Riemannian Gravity

7.1 Introduction

From an experimental point of view, the astonishing experimental success of Einstein's theory is undeniable; in fact, it has successfully passed by highly accurate experiments both in weak and strong field regimes (for a review on experimental studies in gravity, see, for example, [1] and references therein). From a theoretical point of view, GR is perhaps the most elegant physical theory ever—such a claim is not a hyperbole—which is corroborated through its simple geometrical interpretation of the gravitational effects. In this respect, GR background is based on a pseudo-Riemannian geometry in which the metric, $g_{\mu\nu}$, is the only fundamental geometrical entity, while the role of the connection is reduced to a mere testimonial notational appearance. Indeed, the metric entirely determines the structure of the connection whose explicit form in turn is given by the Levi-Civita one ($L^{\alpha}_{\mu\nu}$). As a consequence, to have the Levi-Civita connection as a solution, one must impose metric compatibility and torsion-free conditions leading to the vanishing of the non-metricity and torsion tensors (we shall define these objects later).

Without any empirical evidence that ensures that space-time is described by a pseudo-Riemannian geometry, natural questions arise, namely why not consider extended geometries to describe space-time? What is the most natural manner to generalize the pseudo-Riemannian geometry? The key to answering these questions lies in the treating of non-Riemannian geometries. In particular, the most natural way to generalize the (pseudo)-Riemannian geometry is to relax the non-metricity and torsionless constraints, rendering the non-metricity and torsion tensors non-zeros. As a result, this provides a general connection in a distinguishing way from GR. Equivalently, this description can be thought of as the metric and connection as independent geometrical objects; such an idea was first introduced by Palatini [180]. Gravitational theories defined on this generalized background are often referred to as metric-affine theories of gravity. The richness of such theories lies in the fact that—by assuming torsion and non-metricity as new degrees of freedom or, as the same, taking the metric and connection as independent fields—it potentially allows building new

© The Author(s), under exclusive license to Springer Nature Switzerland AG 2023
A. Petrov et al., *Introduction to Modified Gravity*,
SpringerBriefs in Physics, https://doi.org/10.1007/978-3-031-46634-2_7

effective interaction terms in the action: For example, coupling involving the matter sources with the torsion, as it occurs in Einstein-Cartan theory, or unconventional interactions between a scalar field and higher order curvature invariants, as arises in projective invariant metric-affine Chern-Simons modified theory of gravity.

In this chapter, we review the main mathematical tools of non-Riemannian geometry and give some examples of metric-affine theories of gravity.

7.2 Metric-Affine Geometrical Tools

Our aim here is briefly to define the fundamental geometrical quantities in non-Riemannian geometry. To start with, let us first distinguish the metricity and affinity features. The former is related to the notion of length, angle, distance, etc. thus, locally defined, while the latter is associated with the notion of parallelism or, more generically, to properties invariant under translations, for example, the parallel transport of vectors from one point to another. Hence, both properties are completely independent and should not be confused, except in (pseudo-)Riemannian geometry, where all geometrical entities are entirely constructed from the metric.

We shall now show some useful relations in non-Riemannian geometry. Let us begin with the definition of the non-metricity tensor, i.e.

$$Q_{\alpha\beta\gamma} = \nabla_{\alpha}^{(\Gamma)} g_{\beta\gamma}, \tag{7.1}$$

which clearly recovers the pseudo-Riemannian geometry when imposing the metric-compatibility condition, $Q_{\alpha\beta\gamma} = 0$. Note that our convention for the covariant derivative follows the same one adopted in [181],

$$\nabla_{\mu}^{(\Gamma)} B_{\nu}^{\alpha} = \partial_{\mu} B_{\nu}^{\alpha} + \Gamma^{\alpha}{}_{\mu\lambda} B_{\nu}^{\lambda} - \Gamma^{\lambda}{}_{\nu\mu} B_{\lambda}^{\alpha}. \tag{7.2}$$

Recall that the above affine connection $\Gamma^{\lambda}{}_{\mu\nu}$, in non-Riemannian manifolds, does not coincide with the Levi-Civita one. In fact, this connection can be conveniently split into their symmetric and antisymmetric pieces, i.e.

$$\Gamma^{\lambda}{}_{\mu\nu} = \Gamma^{\lambda}{}_{(\mu\nu)} + \frac{1}{2} T^{\lambda}{}_{\mu\nu}, \tag{7.3}$$

where $T^{\lambda}{}_{\mu\nu} \equiv 2\Gamma^{\lambda}{}_{[\mu\nu]}$ is the torsion tensor. The aforementioned equation can be yet recast as follows:

$$\Gamma^{\lambda}{}_{\mu\nu} = L^{\lambda}{}_{\mu\nu} + D^{\lambda}{}_{\mu\nu} + K^{\lambda}{}_{\mu\nu}, \tag{7.4}$$

where

$$L^{\lambda}{}_{\mu\nu} = \frac{1}{2} g^{\lambda\theta} \left(\partial_{\mu} g_{\nu\theta} + \partial_{\nu} g_{\mu\theta} - \partial_{\theta} g_{\mu\nu} \right) \tag{7.5}$$

is the Levi-Civita connection,

$$D^\lambda{}_{\mu\nu} = \frac{1}{2}\left(Q^\lambda{}_{\mu\nu} - Q_\mu{}^\lambda{}_\nu - Q_\nu{}^\lambda{}_\mu\right) \tag{7.6}$$

is sometimes called the distortion tensor and

$$K^\lambda{}_{\mu\nu} = \frac{1}{2}\left(T^\lambda{}_{\mu\nu} - T_\mu{}^\lambda{}_\nu - T_\nu{}^\lambda{}_\mu\right) \tag{7.7}$$

is the so-called contortion tensor [182]. In practical terms, the distortion and contortion tensors encode the departures from Riemannian geometry. Note that the distortion tensor obeys $D_{\theta\nu\alpha} = D_{\theta\alpha\nu}$ and $D_{[\theta\nu]\alpha} = Q_{[\theta\nu]\alpha}$. While the contortion tensor satisfies $K_{\theta\alpha\nu} = -K_{\alpha\theta\nu}$, the inverse relation between contortion and torsion tensors is simply given by $K^\lambda{}_{\mu\nu} - K^\lambda{}_{\nu\mu} = T^\lambda{}_{\mu\nu}$.

The Riemann tensor is defined as

$$\begin{aligned}
R^\lambda{}_{\alpha\mu\nu}(\Gamma) &= \partial_\mu \Gamma^\lambda{}_{\alpha\nu} - \partial_\nu \Gamma^\lambda{}_{\alpha\mu} + \Gamma^\sigma{}_{\alpha\nu}\Gamma^\lambda{}_{\sigma\mu} - \Gamma^\sigma{}_{\alpha\mu}\Gamma^\lambda{}_{\sigma\nu} \\
&= {}^{(L)}R^\lambda{}_{\alpha\mu\nu} + \nabla^{(L)}_\mu B^\lambda{}_{\alpha\nu} - \nabla^{(L)}_\nu B^\lambda{}_{\alpha\mu} + B^\sigma{}_{\alpha\nu}B^\lambda{}_{\sigma\mu} - B^\sigma{}_{\alpha\mu}B^\lambda{}_{\sigma\nu},
\end{aligned} \tag{7.8}$$

where ${}^{(L)}R^\lambda{}_{\alpha\mu\nu}$ is the usual Riemann tensor of the Levi-Civita connection, $\nabla^{(L)}_\nu$ is the Levi-Civita covariant derivative and $B^\sigma{}_{\alpha\nu} \equiv D^\sigma{}_{\alpha\nu} + K^\sigma{}_{\alpha\nu}$. As a particular case of metric-affine geometry, one can cite the Riemann-Cartan geometry which is featured by the torsion and curvature tensors, whereas the non-metricity is set to be zero. Similarly, one can define the other important geometrical tensors: the Ricci tensor $R_{\mu\nu}$ and the Ricci scalar R, respectively, as

$$\begin{aligned}
R_{\alpha\nu}(\Gamma) \equiv R^\lambda{}_{\alpha\lambda\nu}(\Gamma) &= \partial_\lambda \Gamma^\lambda{}_{\alpha\nu} - \partial_\nu \Gamma^\lambda{}_{\alpha\lambda} + \Gamma^\sigma{}_{\alpha\nu}\Gamma^\lambda{}_{\sigma\lambda} - \Gamma^\sigma{}_{\alpha\lambda}\Gamma^\lambda{}_{\sigma\nu} \\
&= {}^{(L)}R_{\alpha\nu} + \nabla^{(L)}_\lambda B^\lambda{}_{\alpha\nu} - \nabla^{(L)}_\nu B^\lambda{}_{\alpha\lambda} + B^\sigma{}_{\alpha\nu}B^\lambda{}_{\sigma\lambda} - B^\sigma{}_{\alpha\lambda}B^\lambda{}_{\sigma\nu},
\end{aligned} \tag{7.9}$$

$$R(\Gamma) \equiv g^{\mu\nu}R_{\mu\nu} = {}^{(L)}R + \nabla^{(L)}_\lambda B^\lambda - \nabla^{(L)}_\nu B^{\lambda\nu}{}_\lambda + B^\sigma B^\lambda{}_{\sigma\lambda} - B^{\sigma\nu}{}_\lambda B^\lambda{}_{\sigma\nu}, \tag{7.10}$$

where ${}^{(L)}R_{\alpha\nu}$ is the usual Ricci tensor of the Levi-Civita connection and $B^\lambda \equiv B^\lambda{}_{\alpha\nu}g^{\alpha\nu}$.

The curvature, non-metricity and torsion tensors must obey the following generalized Bianchi identities:

$$\nabla^{(\Gamma)}_{[\beta} R^\lambda{}_{|\alpha|\mu\nu]} = R^\lambda{}_{\alpha\sigma[\beta}T^\sigma{}_{\mu\nu]}, \tag{7.11}$$

$$R^\lambda{}_{[\alpha\mu\nu]} = -\nabla^{(\Gamma)}_{[\alpha}T^\lambda{}_{\mu\nu]} - T^\sigma{}_{[\alpha\mu}T^\lambda{}_{\nu]\sigma}. \tag{7.12}$$

In the following section, we shall address how to implement fermions in the non-Riemannian framework, in particular, in the Einstein-Cartan geometry and their quantum effects in the one-loop approximation.

7.3 Fermions in a Riemann-Cartan Space

7.3.1 General Definitions

Our aim in this section is to couple minimally spin-1/2 fermions to gravity and torsion in a Riemann-Cartan space, since the non-metricity is set to be zero; see for example [182–184]. Let us start by defining the Dirac action, i.e.

$$S_D = \int d^4x \sqrt{|g|} \left[\frac{i}{2} e^\mu_a \left(\bar{\Psi} \gamma^a (\nabla^{(\Gamma)}_\mu \Psi) - (\nabla^{(\Gamma)}_\mu \bar{\Psi}) \gamma^a \Psi \right) - m \bar{\Psi} \Psi \right], \qquad (7.13)$$

where γ^a is the flat-space Dirac matrices and e^a_μ is the tetrad, also called vierbein field. Here, we use Latin letters for $SO(1, 3)$ group indices and Greek letters for space-time indices. In addition, the metric of the space-time can be locally defined in terms of a set of orthonormal bases: $\{e_a(x)\}$ and $\{\theta^a(x)\}$,

$$g = \eta_{ab} e^a \otimes e^b = g_{\mu\nu} dx^\mu \otimes dx^\nu, \qquad (7.14)$$

where η_{ab} is the Minkowski metric. Note that the relation $g_{\mu\nu} = e^a_\mu(x) e^b_\nu(x) \eta_{ab}$ is obtained from the duality condition between both frames. As a result, it is straightforward that the flat-space Dirac matrices are connected with those in curved space through the relations $\gamma_\mu = e^a_\mu \gamma_a$ and $\gamma^\mu = e^\mu_a \gamma^a$, which satisfy the Clifford algebra, $\{\gamma^a, \gamma^b\} = 2\eta^{ab}$ and $\{\gamma^\mu, \gamma^\nu\} = 2g^{\mu\nu}$. The covariant derivatives acting on spinors in Eq.(7.13) are defined by

$$\nabla^{(\Gamma)}_\mu \Psi = \partial_\mu \Psi + \Gamma_\mu \Psi, \qquad (7.15)$$

$$\nabla^{(\Gamma)}_\mu \bar{\Psi} = \partial_\mu \bar{\Psi} - \bar{\Psi} \Gamma_\mu, \qquad (7.16)$$

with

$$\Gamma_\mu = \frac{i}{4} \omega_{\mu ab} \sigma^{ab}, \qquad (7.17)$$

where $\omega_{\mu ab}$ are the components of the spin connection whose explicit form is given by

$$\omega_{\mu ab} = \eta_{ac} e^c_\nu \left(\partial_\mu e^\nu_b + e^\beta_b \Gamma^\nu_{\mu\beta} \right), \qquad (7.18)$$

and $\sigma^{ab} = \frac{i}{2} [\gamma^a, \gamma^b]$ are the generators of the covering Lorentz group in the spinor representation.

Using the Cartan equations, one can write the torsion tensor in terms of the vierbein and spin connection, explicitly,

$$T^a_{\mu\nu} = \partial_\mu e^a_\nu - \partial_\nu e^a_\mu + \omega_\mu{}^a{}_\nu - \omega_\nu{}^a{}_\mu. \qquad (7.19)$$

The spin connection can be decomposed into two parts in Riemann-Cartan geometry,

$$\omega_\mu{}^{ab} = \bar{\omega}_\mu{}^{ab} + K^{ba}{}_\mu, \tag{7.20}$$

where $\bar{\omega}_\mu{}^{ab}$ is the torsionless Cartan connection which is entirely determined by the vierbeins. Its explicit form is given by

$$\bar{\omega}_{cab} = -\Omega_{cab} - \Omega_{acb} + \Omega_{bca}, \tag{7.21}$$

where $\Omega_{abc} = \frac{1}{2} e_b{}^\mu e_c{}^\nu \left(\partial_\mu e_{\nu a} - \partial_\nu e_{\mu a} \right)$ are the anholonomy coefficients (also called Ricci rotation coefficients; see, for example, [186]). The second term on the r.h.s. of Eq.(7.20) is precisely the contortion tensor, defined in Eq.(7.7).

It is more convenient to rewrite Eq. (7.13) in a more appropriate way. First, we integrate it by parts and then substitute Eqs. (7.20) and (7.7) into Eq. (7.13) to find

$$\begin{aligned}
S_D &= \int d^4x \sqrt{|g|}\, \bar{\Psi} \left[i\gamma^\mu \bar{\nabla}_\mu + \frac{1}{8} S_\mu \gamma_5 \gamma^\mu - m \right] \Psi \\
&= \bar{S}_D + \frac{1}{8} \int d^4x \sqrt{|g|}\, S_\mu \bar{\Psi} \gamma_5 \gamma^\mu \Psi \\
&= \bar{S}_D - \frac{1}{8} \int d^4x \sqrt{|g|}\, S_\mu J_5^\mu,
\end{aligned} \tag{7.22}$$

with $\bar{S}_D$ representing the usual Dirac action which is purely defined in terms of the Cartan connection and $J_5^\mu \equiv \bar{\Psi} \gamma^\mu \gamma_5 \Psi$ being the axial spin vector current. With these aforementioned definitions, the tilded covariant derivatives acting on spinors are written as follows:

$$\bar{\nabla}_\mu \Psi = \partial_\mu \Psi + \bar{\Gamma}_\mu \Psi, \tag{7.23}$$
$$\bar{\nabla}_\mu \bar{\Psi} = \partial_\mu \bar{\Psi} - \bar{\Psi} \bar{\Gamma}_\mu, \tag{7.24}$$

with $\bar{\Gamma}_\mu$ defined similar to Eq.(7.17) just with $\bar{\omega}_\mu{}^{ab}$ replacing $\omega_\mu{}^{ab}$. Furthermore, we defined the axial-vector torsion $S_\mu = \epsilon_{\mu\nu\alpha\beta} T^{\nu\alpha\beta}$. Hence, just the axial part of torsion interacts with fermions in Riemann-Cartan spaces [182, 185].

We will interpret the axial-vector torsion as a formally external vector in a way similar to what happens in Sect. 3.2, with S_μ playing the role of b_μ, that is, the constant axial-vector breaking the Lorentz symmetry.

7.3.2 One-Loop Induced Nieh-Yan Topological Term

We here intend to compute the one-loop corrections to the spinor effective action. As a first step, let us rewrite the Dirac action as

$$S_D = \int d^4x \, ee_a^\mu \, \bar{\Psi} \left[i\bar{\nabla}_\mu \gamma^a - \frac{1}{8} S_\mu \gamma^a \gamma_5 - m \right] \Psi, \qquad (7.25)$$

so it yields the following one-loop spinor effective action:

$$\Gamma^{(1)} = -i \, \mathrm{Tr} \ln(i\bar{\slashed{\nabla}} - m - \frac{1}{8}\slashed{S}\gamma_5). \qquad (7.26)$$

We compute this effective action up to the first order in derivatives in much the same way as Sect. 3.2.3, see [83, 186] for the step-by-step calculations, with the only difference in replacing b_μ by S_μ. Therefore, the finite contribution for the spinor one-loop effective action is given by

$$\Gamma^{(1)} = \frac{1}{216\pi^2} \int d^4x \sqrt{|g|} \, \epsilon^{\mu\nu\lambda\rho} S_\mu \left(\partial_\nu \bar{\omega}_{\lambda ab} \bar{\omega}_\rho^{\;\;ab} + \frac{2}{3} \bar{\omega}_{\nu ab} \bar{\omega}_\lambda^{\;bc} \bar{\omega}_{\rho c}^{\;\;a} \right) \quad (7.27)$$

$$= \frac{1}{216\pi^2} \int d^4x \sqrt{|g|} S_\mu \bar{C}^\mu, \qquad (7.28)$$

where

$$\bar{C}^\mu = \epsilon^{\mu\nu\lambda\rho} \left[\bar{\omega}_\rho^{\;ba} \partial_\nu \bar{\omega}_{\lambda ab} + \frac{2}{3} \bar{\omega}_{\nu ab} \, \bar{\omega}_\lambda^{\;bc} \, \bar{\omega}_{\rho c}^{\;a} \right] \qquad (7.29)$$

is the topological Chern-Simons vector current defined before as K^μ, in Eq. (3.10), in terms of the affine connection. It should be noted that all divergences in this contribution cancel out similar to the gravitational Chern-Simons term, as expected. It is worth mentioning that the *r.h.s.* of Eq. (7.28) represents a contact term between two topological currents, i.e. Nieh-Yan and Chern-Simons ones. Therefore, we refer to it as to Nieh-Yan term, which is defined by

$$NY = dS, \quad S = e^a \wedge T^a, \qquad (7.30)$$

where NY is the Nieh-Yan 4-form and S is the Nieh-Yan 3-form topological current, which is related to the axial-vector of the torsion tensor; see [186].

In summary, one can observe that the axial-vector of the torsion S_μ plays the role of a Lorentz-breaking vector within the gravitational Chern-Simons term, therefore it is reasonable to conjecture that the non-zero torsion can be treated as a possible origin for the Lorentz symmetry breaking.

7.4   Metric-Affine Bumblebee Model

Bumblebee theories of gravity are models that incorporate spontaneous LSB due to the presence of a dynamical vector field, B_μ, frequently called the bumblebee field, which develops a nontrivial vacuum expectation value (VEV) and then allows for

Lorentz/CPT violations. Therefore, bumblebee models set up a simple framework to investigate interesting features of LSB. Such models have been massively explored in pseudo-Riemannian spaces (see, for example, [108]), though some works propose their non-Riemannian extensions: for example, in [108], where bumblebee models are defined on a Riemann-Cartan manifold. A metric-affine extension of the bumblebee model was first proposed in [187], in which the authors have investigated some classical and quantum aspects of this theory.

The action of the metric-affine bumblebee model is given by [187]

$$S_B = \int d^4x \sqrt{|g|}\left[\frac{1}{2\kappa^2}\left(R(\Gamma) + \xi B^\alpha B^\beta R_{\alpha\beta}(\Gamma)\right) - \frac{1}{4}B^{\mu\nu}B_{\mu\nu} - V(B^\mu B_\mu \pm b^2)\right]$$
$$+ \int d^4x \sqrt{|g|}\mathcal{L}_M(g_{\mu\nu}, \Gamma^\alpha{}_{\mu\nu}, \psi), \tag{7.31}$$

where $\mathcal{L}_M$ is the Lagrangian of matter sources and B_μ is the bumblebee field. As said before, the bumblebee field B_μ acquires a non-zero vacuum expectation value (VEV), say $B_\mu = b_\mu$, which in turn is defined as one of the minima of the potential V, i.e. $V'(B^\mu B_\mu \pm b^2) = 0$; the prime stands for the derivative with respect to the argument of V. Within this framework, any observable that couples with b_μ gets a preferred frame in space-time, then the Lorentz symmetry is broken spontaneously. It is worth calling attention we are referring to local Lorentz symmetry breaking since we are dealing with curved spaces. Furthermore, $B_{\mu\nu}$ is the bumblebee strength field associated with B_μ. Its explicit form is consistently defined as follows:

$$B_{\mu\nu} = (dB)_{\mu\nu}. \tag{7.32}$$

Note that the antisymmetric part of the Ricci tensor does not develop a nontrivial contribution in the second term of (4.1). As a consequence, it does not contribute to dynamical equations for the torsion.

7.4.1 Field Equations

Varying the action 7.31 with respect to $g_{\mu\nu}$, $\Gamma^\alpha{}_{\mu\nu}$ and B_μ, one finds their respective field equations, namely

$$R_{(\mu\nu)}(\Gamma) - \frac{1}{2}g_{\mu\nu}\left(R(\Gamma) + \xi B^\alpha B^\beta R_{\alpha\beta}(\Gamma)\right) + 2\xi\left(B_{(\mu}R_{\nu)\beta}(\Gamma)\right)B^\beta = \kappa^2(T^M_{\mu\nu} + T^B_{\mu\nu}),$$
$$\tag{7.33}$$
$$\nabla^{(\Gamma)}_\lambda\left[\sqrt{|h|}h^{\nu\mu}\right] - \delta^\mu_\lambda\nabla^{(\Gamma)}_\rho\left[\sqrt{|h|}h^{\nu\rho}\right] = \sqrt{|h|}\left[T^\mu{}_{\lambda\alpha}h^{\nu\alpha} + T^\alpha{}_{\alpha\lambda}h^{\nu\mu} - \delta^\mu_\lambda T^\alpha{}_{\alpha\beta}h^{\nu\beta}\right]$$
$$+ \kappa^2\Delta_\lambda{}^{\mu\nu}, \tag{7.34}$$
$$\nabla^{(g)}_\mu B^{\mu\nu} = -\frac{\xi}{\kappa^2}g^{\nu\alpha}B^\beta R_{\alpha\beta}(\Gamma) + 2V'B^\nu, \tag{7.35}$$

where the total stress-energy tensor has been split into two pieces: one, coming from contributions of the matter sources ($T^M_{\mu\nu}$) and, the other, of the bumblebee field ($T^B_{\mu\nu}$). They are explicitly defined as follows:

$$T^M_{\mu\nu} = -\frac{2}{\sqrt{|g|}}\frac{\delta(\sqrt{|g|}\mathcal{L}_M)}{\delta g^{\mu\nu}}, \tag{7.36}$$

$$T^B_{\mu\nu} = B_{\mu\sigma}B_\nu{}^\sigma - \frac{1}{4}g_{\mu\nu}B^\alpha{}_\sigma B^\sigma{}_\alpha - Vg_{\mu\nu} + 2V'B_\mu B_\nu. \tag{7.37}$$

We have also defined the auxiliary metric $h^{\mu\nu}$ as

$$\sqrt{|h|}h^{\mu\nu} \equiv \sqrt{|g|}(g^{\mu\nu} + \xi B^\mu B^\nu), \tag{7.38}$$

and

$$\Delta^{\mu\nu}_\lambda \equiv 2\frac{\delta(\sqrt{|g|}\mathcal{L}_M)}{\delta\Gamma^\lambda_{\mu\nu}} \tag{7.39}$$

is the hypermomentum tensor representing the coupling between matter and connection at the level of the field equations.

Proceeding with some manipulations in the metric Eq. (7.33), we are able to decouple the field equations. To see that, by taking the trace of Eq. (7.33), one gets

$$R(\Gamma) = -\kappa^2 T, \tag{7.40}$$

where $T = g^{\mu\nu}T_{\mu\nu}$. Similarly, it is easy to see that

$$B^\mu B^\nu R_{\mu\nu}(\Gamma) = \frac{\kappa^2}{2+3\xi X}\left(-TX + 2B^\mu B^\nu T_{\mu\nu}\right), \tag{7.41}$$

$$B^\mu R_{\mu\nu}(\Gamma) = \frac{\kappa^2}{1+\xi X}\left(B^\mu T_{\mu\nu} - \frac{\xi B_\nu B^\alpha B^\beta T_{\alpha\beta} + B_\nu[1+\xi X]T}{2+3\xi X}\right), \tag{7.42}$$

where $X \equiv g^{\mu\nu}B_\mu B_\nu$. Substituting Eqs. (7.40–7.42) into Eq. (7.33), we arrive at

$$\begin{aligned}
R_{(\mu\nu)}(\Gamma) = \frac{1}{2}\kappa^2 g_{\mu\nu}&\left[\frac{\xi}{2+3\xi X}\left(2B^\alpha B^\beta T_{\alpha\beta} - TX\right) - T\right] \\
&- \frac{\xi\kappa^2}{1+\xi X}\left[B^\alpha\left(B_\mu T_{\alpha\nu} + B_\nu T_{\alpha\mu}\right)\right. \\
&\left.- \frac{B_\mu B_\nu}{2+3\xi X}\left(2(1+\xi X)T + 2\xi B^\alpha B^\beta T_{\alpha\beta}\right)\right] + \kappa^2 T_{\mu\nu},
\end{aligned} \tag{7.43}$$

making it clear that the Ricci tensor is a function of B_μ, $g_{\mu\nu}$ and the matter sources. Consequently, the non-minimal term $B^\mu B^\nu R_{\mu\nu}(\Gamma)$ that appears in the action of the model can be rewritten as a series of unconventional terms on-shell, namely bumblebee's self-interaction ones, involving couplings between the bumblebee and matter sources, and so on.

Let us now turn our attention to the connection Eq. (7.34). One can check that for the bumblebee theory (see [188] for more detail), the connection can be expressed as

$$\Gamma^{\alpha}_{\mu\nu} = {}^{h}\Gamma^{\alpha}_{\mu\nu} + \Upsilon^{\alpha}_{\mu\nu}, \tag{7.44}$$

where

$$^{h}\Gamma^{\alpha}_{\mu\nu} = \frac{1}{2}h^{\alpha\lambda}\left(-\partial_{\lambda}h_{\mu\nu} + \partial_{\mu}h_{\nu\lambda} + \partial_{\nu}h_{\mu\lambda}\right) \tag{7.45}$$

are the Christoffel symbols of the auxiliary metric $h_{\mu\nu}$ and $\Upsilon^{\lambda}_{\mu\nu}$ must satisfy

$$\Upsilon^{\alpha}_{\mu\nu}\left(\delta^{\kappa}_{\alpha}\delta^{\mu}_{\beta}\delta^{\nu}_{\gamma} + \frac{1}{2}\delta^{\mu}_{\alpha}\left(h^{\nu\kappa}h_{\beta\gamma} - \delta^{\nu}_{\beta}\delta^{\kappa}_{\gamma} - \delta^{\nu}_{\gamma}\delta^{\kappa}_{\beta}\right)\right) = \frac{\kappa^{2}}{2\sqrt{|h|}}h^{\kappa\lambda}\left(\Delta_{\beta\gamma\lambda} + \Delta_{\gamma\lambda\beta} + \Delta_{\lambda\beta\gamma}\right) \tag{7.46}$$

in accordance with Eq. (7.34). Recalling that the metric-affine bumblebee gravity possesses projective symmetry, as a result, the projective modes are totally irrelevant. It is realized from Eq. (7.46) that Υ is zero for minimally coupled bosonic fields since the hypermomentum contribution vanishes. However, the situation changes for minimally coupled fermionic fields since now the hypermomentum contribution is nontrivial, which entails a non-zero Υ.

Gett back to the auxiliary metric h appearing in Eq. (7.38), which is convenient to set in a matrix form, i.e.

$$\sqrt{|h|}\hat{h}^{-1} = \sqrt{|g|}\hat{g}^{-1}\left(\hat{I} + \xi\hat{B}B\right), \tag{7.47}$$

where $\hat{h}^{-1}$ is the matrix form for $h^{\mu\nu}$ and $\hat{h}$ is used for the $h_{\mu\nu}$ matrix. Taking the determinant of the previous equation, we find

$$\hat{h} = \hat{g}\det\left(\hat{I} + \xi\hat{B}B\right), \tag{7.48}$$

which, upon insertion into Eq. (7.38), results in

$$h^{\mu\nu} = \frac{1}{\sqrt{\det\left(\hat{I} + \xi\hat{B}B\right)}}g^{\mu\alpha}(\delta^{\nu}_{\alpha} + \xi B^{\nu}B_{\alpha}). \tag{7.49}$$

It is easy to see that

$$h_{\mu\nu} = \sqrt{\det\left(\hat{I} + \xi\hat{B}B\right)}\left[g_{\mu\alpha}\left(\delta^{\alpha}_{\nu} - \frac{\xi}{\det\left(\hat{I} + \xi\hat{B}B\right)}B^{\alpha}B_{\nu}\right)\right], \tag{7.50}$$

where we have used the relation $h^{\mu\nu}h_{\alpha\nu} = \delta^\mu_\alpha$. Taking into account the identity $\det(\hat{I} + \xi \hat{B}B) = 1 + \xi X$, Eqs. (7.49–7.50) can be rewritten as

$$h^{\mu\nu} = \frac{1}{\sqrt{1+\xi X}} g^{\mu\alpha}(\delta^\nu_\alpha + \xi B^\nu B_\alpha), \tag{7.51}$$

$$h_{\mu\nu} = \sqrt{1+\xi X}\left[g_{\mu\alpha}\left(\delta^\alpha_\nu - \frac{\xi}{1+\xi X} B^\alpha B_\nu\right)\right]. \tag{7.52}$$

Notably, one can see that $h_{\mu\nu}$ is related with $g_{\mu\nu}$ and B_μ. In the literature, such relations are commonly referred to as disformal transformations [190]. In Lorentz-violating scenarios, similar metric structures have been considered even in Riemannian spaces (see, for example, [191]).

Now, we turn our attention to the dynamical bumblebee equation. Let us start by substituting Eq. (7.42) into Eq. (7.35), then one finds that its dependence on the connection is eliminated and, thus, we have a direct relation between $g_{\mu\nu}$ and b_μ. By doing so, the remaining equation is a Proca-like one in the curved space-time,

$$\nabla^{(g)}_\alpha B^{\alpha\mu} = \mathcal{M}^\mu_{\ \nu} B^\nu, \tag{7.53}$$

where

$$\mathcal{M}^\mu_{\ \nu} = \left(\frac{\xi T}{2+3\xi X} + \frac{\xi^2 B^\alpha B^\beta T_{\alpha\beta}}{(1+\xi X)(2+3\xi X)} + 2V'\right)\delta^\mu_{\ \nu} - \frac{\xi}{(1+\xi X)} T^{\mu\alpha} g_{\nu\alpha} \tag{7.54}$$

is defined as the effective mass-squared tensor.

Actually, Eq. (7.53) displays new unconventional couplings between the bumblebee field and the energy-momentum tensor without any similarity to the metric case. Note that the bumblebee field equation can present instabilities. In fact, as long as the determinant of the mass-squared tensor assumes negative values (it is completely possible since there exists an opposite sign between both terms in Eq. (7.54)), the bumblebee field describes a tachyonic particle.

The divergence of Eq. (7.53) leads to the following constraint:

$$\nabla^{(g)}_\nu (m^2_{eff} B^\nu + J^\nu_{eff}) = 0, \tag{7.55}$$

where

$$m^2_{eff} = \frac{\xi T}{2+3\xi X} + \frac{\xi^2 B^\alpha B^\beta T_{\alpha\beta}}{(1+\xi X)(2+3\xi X)} + 2V';$$

$$J^\nu_{eff} = -\frac{\xi}{(1+\xi X)} T^{\nu\alpha} B_\alpha. \tag{7.56}$$

Indeed, the constraint equation can be interpreted as a conservation. In order to see that, we just need to proceed with the redefinition $\bar{J}^{\nu}_{eff} = m^2_{eff} B^{\nu} + J^{\nu}_{eff}$ and then substitute it in the constraint equation to find $\nabla^{(g)}_{\nu} \bar{J}^{\nu}_{eff} = 0$.

7.4.2 Spinor Sector and Spontaneous Lorentz Symmetry Breaking

The Dirac action in metric-affine theories minimally coupled to the geometry has the same form as that defined in a Riemann-Cartan space, Eq. (7.13), with the difference that the non-metricity is no longer set to be zero. In this respect, if the frame $\{e_a(x)\}$ and its co-frame $\{\theta^a(x)\}$ are related to the metric $g_{\mu\nu}$ by means of Eq. (7.14), we can locally define another set of orthonormal (co-)frames: $\{E_a(x)\}$ and $\{\Theta^a(x)\}$, which are related to the auxiliary metric $h_{\mu\nu}$ by the following relation:

$$h = \eta_{ab}\Theta^a \otimes \Theta^b = h_{\mu\nu}dx^{\mu} \otimes dx^{\nu}, \tag{7.57}$$

where $h_{\mu\nu} = E_{\mu}{}^a(x)E_{\nu}{}^b(x)\eta_{ab}$. Under these considerations and using Eq. (7.38), one has the relation between both vierbeins, i.e.

$$e^a{}_{\mu} = (1+\xi X)^{-1/4} E^a_{\lambda} \left[\delta^{\lambda}{}_{\mu} - \frac{(1-\sqrt{1+\xi X})}{X(1+\xi X)^{1/2}} \tilde{B}^{\lambda} B_{\mu} \right]; \tag{7.58}$$

$$e^{\mu}{}_a = (1+\xi X)^{1/4} E^{\lambda}{}_a \left[\delta^{\mu}{}_{\lambda} + \frac{(1-\sqrt{1+\xi X})}{X(1+\xi X)} \tilde{B}^{\mu} B_{\lambda} \right], \tag{7.59}$$

where we have denoted tensorial quantities, which have been raised or lowered with the auxiliary metric, $h_{\mu\nu}$, with a tilde. Note that, similar to what happens in the untilded frame, we can define $\tilde{\gamma}^{\mu} = E^{\mu}{}_a \gamma^a$ which must fulfill $\{\tilde{\gamma}^{\mu}, \tilde{\gamma}^{\nu}\} = 2h^{\mu\nu}$. Putting all the above information together, we are able to find an analytical expression for the hypermomentum and $\Upsilon^{\alpha}_{\mu\nu}$ due to Dirac fields; they are explicitly given by (see [188] for the detailed computation)

$$\Delta^{\mu\nu}_{\lambda} = -\frac{\sqrt{|h|}}{2} \left(E^{\mu}_a + (1+\xi X)^{1/4} \theta^{\mu}_a \right) \epsilon^{abcd} J_{5d} \left[\eta_{bk} E^k{}_{\lambda} E^{\nu}{}_c \right.$$
$$\left. + (1+\xi X)^{-1/4} \eta_{bk} E^c{}_{\lambda} \theta^{\nu}{}_c - \theta_{b\lambda} \left((1+\xi X)^{1/4} E^{\nu}{}_c + \theta^{\nu}{}_c \right) \right], \tag{7.60}$$

$$\Upsilon^{\kappa}_{\beta\gamma} = \kappa^2 \epsilon^{abcd} J_{5d} h^{\kappa\lambda} \Bigg[\left(h_{\gamma\mu} E^{\mu}_a + \mathfrak{X}^{1/2} \theta_{\gamma a} \right) \left[\eta_{bk} E^k_{\ \beta} h_{\lambda\nu} E^{\nu}_{\ c} + \mathfrak{X}^{-1/2} \eta_{bk} E^c_{\ \beta} \theta_{\lambda c} \right.$$

$$\left. - \theta_{b\beta} \left(\mathfrak{X}^{1/2} h_{\lambda\nu} E^{\nu}_{\ c} + \theta_{\lambda c} \right) \right]$$

$$+ \left(h_{\lambda\mu} E^{\mu}_a + \mathfrak{X}^{1/2} \theta_{\lambda a} \right) \left[\eta_{bk} E^k_{\ \gamma} h_{\beta\nu} E^{\nu}_{\ c} + \mathfrak{X}^{-1/2} \eta_{bk} E^c_{\ \gamma} \theta_{\beta c} \right.$$

$$\left. - \theta_{b\gamma} \left(\mathfrak{X}^{1/2} h_{\beta\nu} E^{\nu}_{\ c} + \theta_{\beta c} \right) \right]$$

$$+ \left(h_{\beta\mu} E^{\mu}_a + \mathfrak{X}^{1/2} \theta_{\beta a} \right) \left[\eta_{bk} E^k_{\ \lambda} h_{\gamma\nu} E^{\nu}_{\ c} + \mathfrak{X}^{-1/2} \eta_{bk} E^c_{\ \lambda} \theta_{\gamma c} \right.$$

$$\left. - \theta_{b\lambda} \left(\mathfrak{X}^{1/2} h_{\gamma\nu} E^{\nu}_{\ c} + \theta_{\gamma c} \right) \right] \Bigg], \tag{7.61}$$

where we have defined the following quantities for convenience: $\mathfrak{X} = (1 + \xi X)^{1/2}$, $\theta^a_b = B^a B_b$ and $J_5^a = \bar{\Psi} \gamma^a \gamma_5 \Psi$ is the axial fermionic current. In this scenario, upon integration of the $g_{\mu\nu}$ and e^{μ}_a, we obtain the Einstein-frame representation of the Dirac action 7.13, i.e.

$$S_D = \int d^4 x \sqrt{|h|} \left\{ \bar{\Psi} \left(i \tilde{\Gamma}^{\mu} \tilde{\nabla}_{\mu} - M \right) \Psi - \kappa^2 \Sigma(h, B, J_5) \right\} \tag{7.62}$$

where $\tilde{\nabla} \equiv \nabla^{(h)}$ acts on spinors like Eqs. (7.23, 7.24) just with $\tilde{\omega}_{\mu}^{\ ab}$ replacing $\bar{\omega}_{\mu}^{\ ab}$, where

$$\tilde{\omega}_{cab} = -\tilde{\Omega}_{cab} - \tilde{\Omega}_{acb} + \tilde{\Omega}_{bca}, \tag{7.63}$$

with $\tilde{\Omega}_{abc} = \frac{1}{2} E^{\mu}_b E^{\nu}_c \left(\partial_{\mu} E_{\nu a} - \partial_{\nu} E_{\mu a} \right)$. The operators Γ^{μ} and M are expanded in the basis of 16 Dirac matrices in the spinor space of the Clifford algebra. Explicitly,

$$\tilde{\Gamma}^{\mu} = \tilde{\gamma}^{\mu} + c^{\mu}_{\alpha} \tilde{\gamma}^{\alpha} + d^{\mu}_{\alpha} \gamma_5 \tilde{\gamma}^{\alpha} + e^{\mu} I + i f^{\mu} \gamma_5 + \frac{1}{2} g^{\mu\lambda\alpha} \sigma_{\lambda\alpha},$$

$$M = m_0 + i m_5 \gamma_5 + a_{\mu} \tilde{\gamma}^{\mu} + k_{\mu} \tilde{\gamma}^{\mu} \gamma_5 + \frac{1}{2} H^{\mu\nu} \sigma_{\mu\nu}, \tag{7.64}$$

where $c^{\mu}_{\alpha}, d^{\mu}_{\alpha}, e^{\mu}, f^{\mu}, g^{\mu\lambda\alpha}, m_5, a_{\mu}, k_{\mu}$ and $H^{\mu\nu}$ are Lorentz- and/or CPT-dynamical violating coefficients, and we have defined $m_0 = \frac{m}{(1+\xi X)^{1/4}}$. Note, however, that the only non-zero coefficients are

$$c^{\mu}_{\alpha} = \xi^{(1)} B_{\alpha} \tilde{B}^{\mu};$$

$$a_{\mu} = i \xi^{(2)} \tilde{T}_{\mu}; \tag{7.65}$$

$$k_{\mu} = -\xi^{(2)} \tilde{S}_{\mu},$$

with

$$\xi^{(1)} = \frac{(1 - \sqrt{1 + \xi X})}{X(1 + \xi X)};$$

$$\xi^{(2)} = \frac{(1 - \sqrt{1 + \xi X})}{2X(1 + \xi X)^{3/4}};$$

$$\tilde{T}_\mu = (1 + \xi X)^{-1/4} E^a{}_\mu \left(\nabla^h_\nu \tilde{B}^\nu B_a \right),$$

$$\tilde{S}_\mu = \frac{1}{2(1 + \xi X)^{1/2}} \epsilon^{abcd} E_{\mu d} E^\lambda{}_b \tilde{B}_c \left[\left(1 - \sqrt{1 + \xi X} \right) E^\nu_a \nabla^h_\lambda \tilde{B}_\nu + B_\nu \nabla^h_\lambda E^\nu{}_a \right].$$

$$(7.66)$$

The constant couplings $\xi^{(i)}$ previously defined are nonlinearly dependent on ξ and X. In particular, as long as we restrict our analysis to small ξ situation, the $\xi^{(i)}$ no longer depend on X, but only linearly on ξ.

7.4.3 One-Loop Divergent Contributions to the Fermionic Effective Action

We now deal with quantum corrections of the fermionic matter minimally coupled with the metric and connection within the metric-affine bumblebee gravity. In order to investigate the effects of Lorentz symmetry breaking at the quantum level, it is more instructive to evaluate the fermionic one-loop level in the Einstein frame. So, in this case, the spinor effective action is given by

$$\Gamma^{(1)}[h_{\mu\nu}, B_\mu] = -i \ln \det \Delta = -i \operatorname{Tr} \ln \Delta, \tag{7.67}$$

where

$$\begin{aligned}
\Delta[h_{\mu\nu}, B_\mu] &= i \tilde{\Gamma}^\mu \tilde{\nabla}_\mu - M \\
&= i \tilde{\gamma}^\mu \tilde{\nabla}_\mu + i \frac{(1 - \sqrt{1 + \xi X})}{X(1 + \xi X)} (\tilde{\gamma}^\alpha B_\alpha) \tilde{B}^\mu \tilde{\nabla}_\mu - i \frac{(1 - \sqrt{1 + \xi X})}{X(1 + \xi X)^{3/4}} \tilde{\gamma}^\mu \tilde{T}_\mu \\
&\quad + \frac{(1 - \sqrt{1 + \xi X})}{2X(1 + \xi X)^{3/4}} \tilde{\gamma}^\mu \gamma_5 \tilde{S}_\mu - \frac{m}{(1 + \xi X)^{1/4}}
\end{aligned} \tag{7.68}$$

is the modified Dirac operator. As usual, the aforementioned one-loop effective action is divergent, thereby a regularization scheme is required in order to separate the finite contribution from the divergent one. Furthermore, since the effective action possesses non-minimal couplings between the bumblebee field and the covariant derivative, it becomes necessary to use the Barvinsky-Vilkovisky method [189].

Note that the fermionic one-loop action can be rewritten as ([188])

$$\Gamma^{(1)}[h_{\mu\nu}, B_\mu] = -\frac{i}{2} \ln \det \hat{H} = -\frac{i}{2} \operatorname{Tr} \ln \hat{H}, \tag{7.69}$$

where

$$\hat{H} = \left(i\tilde{\Gamma}^\mu\tilde{\nabla}_\mu - M\right)\left(i\tilde{\Gamma}^\mu\tilde{\nabla}_\mu + M^*\right)$$
$$= -\tilde{\Gamma}^\mu\tilde{\Gamma}^\nu\tilde{\nabla}_\mu\tilde{\nabla}_\nu - \tilde{\Gamma}^\mu(\tilde{\nabla}_\mu\tilde{\Gamma}^\nu)\tilde{\nabla}_\nu + i\tilde{\Gamma}^\mu\tilde{\nabla}_\mu M^* + i\tilde{\Gamma}^\mu M^*\tilde{\nabla}_\mu - iM\tilde{\Gamma}^\mu\tilde{\nabla}_\mu$$
$$- MM^*.$$

$$(7.70)$$

Once we are focusing on the perturbative aspects, we restrict ourselves to the weak-coupling regime, $\xi \ll 1$. In this scenario, upon proceeding with any manipulation, the quadratic operator takes the form

$$\hat{H} = -\left(\tilde{\Box}\hat{1} + \hat{H}^{\mu\nu}\tilde{\nabla}_\mu\tilde{\nabla}_\nu + 2\hat{X}^\mu\tilde{\nabla}_\mu + \hat{\Pi}\right), \qquad (7.71)$$

with

$$\tilde{\Box} = h^{\mu\nu}\tilde{\nabla}_\mu\tilde{\nabla}_\nu,$$
$$\hat{H}^{\mu\nu} = -\xi B^\mu B^\nu\hat{1} + \mathcal{O}(\xi^2),$$
$$\hat{X}^\mu = \frac{1}{4}\xi\tilde{T}^\mu\hat{1} - \frac{1}{4}\xi\tilde{\sigma}^{\mu\nu}\gamma_5\tilde{S}_\nu - \frac{1}{4}\xi\tilde{\gamma}^\nu\tilde{\gamma}^\beta\tilde{\nabla}_\nu(B_\beta\tilde{B}^\mu) + \mathcal{O}(\xi^2),$$
$$\hat{\Pi} = \frac{1}{4}\xi\left(\tilde{\nabla}^\mu\tilde{T}_\mu\right)\hat{1} + \frac{1}{8}\xi\tilde{\gamma}^\mu\tilde{\gamma}^\nu\tilde{F}_{\mu\nu} + \frac{i}{4}\xi\gamma_5\tilde{\nabla}^\mu\tilde{S}_\mu + \frac{i}{8}\xi\tilde{\gamma}^\mu\tilde{\gamma}^\nu\gamma_5\tilde{S}_{\mu\nu} - \frac{1}{4}\tilde{R}\hat{1} + m_0^2\hat{1}$$
$$+ \xi B^\nu B_\alpha\tilde{R}_{\nu\mu}\tilde{\gamma}^\alpha\tilde{\gamma}^\mu + \mathcal{O}(\xi^2),$$

$$(7.72)$$

and where we have defined

$$\tilde{F}_{\mu\nu} = \tilde{\nabla}_\mu\tilde{T}_\nu - \tilde{\nabla}_\nu\tilde{T}_\mu, \quad \tilde{S}_{\mu\nu} = \tilde{\nabla}_\mu\tilde{S}_\nu - \tilde{\nabla}_\nu\tilde{S}_\mu, \qquad (7.73)$$

expanded up to the first order in ξ. With this at hand, we are able to compute the divergent piece of the fermionic effective one-loop action by means of the Barvinsky-Vilkovisky method (see [188] for the detailed computation), i.e.

$$\Gamma^{(1)}\Big|_{\text{div}} = \frac{\mu^{D-4}}{(4\pi)^2\epsilon}\int d^D x\sqrt{|h|}\Bigg\{\left(1 - \frac{\xi X}{4}\right)\left[\frac{1}{72}\tilde{R}^2 - \frac{7}{360}\tilde{R}_{\mu\nu\alpha\beta}\tilde{R}^{\mu\nu\alpha\beta} - \frac{1}{45}\tilde{R}_{\mu\nu}\tilde{R}^{\mu\nu}\right]$$
$$+ 2m^4 - \frac{1}{3}m^2\tilde{R} + \frac{1}{12}\xi\tilde{R}\tilde{\nabla}_\mu\tilde{\nabla}_\nu\left(B^\mu B^\nu\right) - m_0^2\xi\tilde{\nabla}_\mu\tilde{\nabla}_\nu\left(B^\mu B^\nu\right) - 2m^4\xi X$$
$$+ 4m^2\xi B^\mu B^\nu\tilde{R}_{\mu\nu} - \frac{4}{9}\xi B^\mu B^\nu\tilde{R}\tilde{R}_{\mu\nu} + \frac{1}{12}\xi\tilde{R}^{\mu\nu\theta\sigma}\tilde{\nabla}_\mu\tilde{\nabla}_\sigma\left(B_\theta B_\nu\right)$$
$$- \frac{7}{30}\xi B^\nu B^\nu\tilde{\nabla}_\mu\tilde{\nabla}_\nu\tilde{R} - \frac{2}{45}\xi B^\mu B^\nu\tilde{R}_{\mu\alpha}\tilde{R}_\nu^\alpha + \frac{1}{30}\xi B^\mu B^\nu\tilde{\Box}\tilde{R}_{\mu\nu} - \frac{1}{6}\xi X\tilde{R}^2$$
$$+ \frac{1}{20}\xi X\tilde{\Box}\tilde{R} + \frac{4}{45}\xi B^\mu B^\nu\tilde{R}^{\alpha\beta}\tilde{R}_{\alpha\mu\beta\nu}\Bigg\} \qquad (7.74)$$

where the boundary terms were dropped. Note that the first three contributions in the above equation mean the usual purely higher curvature terms, while the other unconventional contributions come from the non-minimal coupling between the bumblebee and the curvature.

7.5 Conclusions

In this chapter, we discussed some aspects of non-Riemannian geometries. In particular, we addressed classical and quantum aspects of fermions minimally coupled with gravity in two different non-Riemannian backgrounds, namely Einstein-Cartan and metric-affine geometries. In the former, it is well known that only the axial-vector torsion piece (S_μ) couples with the axial fermionic vector current. Interestingly, S_μ can be formally interpreted as a constant external vector; then, in the light of Lorentz symmetry breaking, any observable, that couples with it, gets Lorentz violation. In this environment, the contribution to the one-loop fermionic effective action is finite and yields a contact term between both topological currents: the Nieh-Yan and Chern-Simons ones. In the latter, we studied fermions minimally coupled with gravity within a metric-affine extension of the bumblebee gravity. This novel model admits an Einstein-frame representation, in which the fermionic sector is now described by a generalized nonlinear Dirac action, thus incorporating a series of unconventional interacting terms between the bumblebee and spinors. At the quantum level, such new terms contribute nontrivially to the divergent piece of the fermionic one-loop effective action, thus suggesting the inclusion of new counterterms in the action involving the bumblebee coupled with the curvature in the Einstein frame.

Certainly, other issues related to modified gravity models formulated on non-Riemannian manifolds must be addressed. Within this context, it is especially interesting, first, to explore in more detail the recently proposed metric-affine version of the Chern-Simons modified gravity [192] (originally claimed in [193]); this particular model sets up as an important ingredient to explore parity-violating effects in curved spaces [194], and, second, to go on exploring metric-affine extensions of Lorentz-breaking gravity models and then investigating the relationship between the non-metricity/torsion and Lorentz symmetry breaking in a way similar to what happens in the metric-affine bumblebee model [187, 188], where the bumblebee VEV was shown to source the non-metricity. Many other relevant studies of the metric-affine gravity can be performed as well.

Chapter 8
Summary

In this book, we discussed various modifications of gravity introduced within the framework of the metric formalism, and, besides this, presented some comments on non-Riemannian gravity models. As we noted, in principle there are two fundamental problems to be solved by the desired modifications of gravity: first, explanation of the cosmic acceleration, and, second, development of a theory consistent from a quantum viewpoint. Within the models we presented, different attempts to solve these problems are taken. It turns out that the problem of cosmic acceleration is solved by many extensions of gravity, and actually the main issue in this context consists in finding the theory fitting better the observational results (for a discussion of cosmological constraining of gravitational models, see, for example, [195] and many other papers). At the same time, the problem of formulating a perturbatively consistent gravity theory appears to be much more complicated. While the simplest way to construct the renormalizable gravity model is based on introducing higher derivative terms, this manner suffers from the problem of arising ghost states. To solve the problem of ghosts, one can follow two ways: either break Lorentz symmetry in a strong manner introducing the HL gravity (effectively it means that we have a higher derivative regularization in a spatial sector only) paying a price of arising a very complicated theory, and moreover, treating the Lorentz symmetry as an essentially low-energy phenomenon, or introduce nonlocality which allows to achieve renormalizability or even to rule out divergences, but, in this case, solving the problems of unitarity and causality would require special efforts.

One more way to solve the problem of renormalizability of gravity is based on its supersymmetric extension. As is well known, a supersymmetric extension of any field theory improves essentially its ultraviolet behavior since so-called "miraculous cancelations" of UV divergences occur [196]. It is well known that the mechanism of these cancelation is very simple—since fermionic contributions carry an extra minus sign, under an appropriate relation between coupling constants occurring due to the supersymmetry, some of fermionic divergent contributions cancel bosonic divergent contributions (for example, while the ϕ^4 theory and Yukawa model display quadratic

A. Petrov et al., *Introduction to Modified Gravity*,
SpringerBriefs in Physics, https://doi.org/10.1007/978-3-031-46634-2_8

divergences, the Wess-Zumino model involving these theories as ingredients displays only logarithmic divergences). Moreover, there are known examples of completely finite supersymmetric theories; the paradigmatic example is the $\mathcal{N} = 4$ super-Yang-Mills theory, where $\mathcal{N}$ is a number of supersymmetries (number of sets of generators of supersymmetry). Clearly, this called interest to a possible supersymmetric extension of gravity, so, the supergravity (SUGRA) was introduced (see [197] for a review). However, the $\mathcal{N} = 1$ SUGRA is still non-renormalizable; therefore, the extensions of SUGRA with larger values of $\mathcal{N}$ began to be introduced. The maximal $\mathcal{N}$ allowing for a consistent theory is 8, for SUGRA (for larger values of $\mathcal{N}$, higher spin fields arise, and they cannot be consistently coupled to gravity). It should be noted also that the interest to SUGRA models with high $\mathcal{N}$ is motivated also by possible applications of these theories to superstrings.

So, let us briefly review the most important results found within $\mathcal{N} = 8$ SUGRA obtained in the series of papers by Bern, Dixon, Kosower and collaborators. In [198] it was proved that the degree of divergence, at $\mathcal{N} = 8$, in D dimensions and L loops, is

$$\omega = (D - 2)L - 10. \tag{8.1}$$

So we see that divergences in four dimensions can begin only from five-loop order! It is interesting to note that the approach from the same paper allows to show that the $\mathcal{N} = 4$ super-Yang-Mills theory is all-loop finite.

Further, on the base of the unitarity cuts approach, in [199], it has been proved that the four-point functions in $\mathcal{N} = 8$ SUGRA satisfy the same finiteness condition in the D-dimensional space-time

$$D < \frac{6}{L} + 4, \tag{8.2}$$

which for $D = 4$ implies all-loop finiteness of these functions. Then, in [200], with the use of some identities applied for sets of more than 30 supergraphs, it was proved that some extra cancelations occur, so, the finiteness of $\mathcal{N} = 8$ SUGRA is achieved up to four loops at $D \leq 5$. Afterwards, in [201] it was proved that the five-loop correction in this theory begins to diverge at $D \geq 24/5$, so, in the four-dimensional space-time, the theory is five-loop finite. Taking all together, we conclude that there is a natural hope that $\mathcal{N} = 8$ SUGRA is all-loop finite in $D = 4$. The next problem consists in extracting some observable results for SUGRA (scattering amplitudes, corrections to GR, etc.) while, up to now, there are only some isolated conclusions.

We conclude our book with the statements that, first, in the study of gravity one still has more questions than answers, and, second, apparently the most promising extensions of gravity are the SUGRA, the nonlocal gravity and the HL gravity. However, each of these modifications still has its difficulties which need to be solved. In principle, there are some other approaches to gravity, for example, treating the gravity as an emergent phenomenon caused by essentially quantum effects [202], asymptotic safety also known as non-perturbative renormalizability, which allows to treat many divergences as nonphysical ones [203], bimetric gravity based on the use of the additional second-rank symmetric tensor and some other approaches. To finish, we note again that in gravity there are still much more questions than answers.

References

1. Q.G. Bailey, arXiv:2305.06325 [gr-qc]
2. V.M. Ponomarev, A.O. Barvinski, Yu.N. Obukhov, *Gauge Approach and Quantization Methods in Gravity Theory* (Nauka, Moscow, 2017)
3. L. Ryder, *Introduction to General Relativity* (Cambridge University Press, 2009)
4. K. Gödel, Rev. Mod. Phys. **21**, 447 (1949)
5. M.J. Rebouças, J. Tiomno, Phys. Rev. D **28**, 1251 (1983)
6. G. 't Hooft and M. J. G. Veltman, Ann. Inst. H. Poincare Phys. Theor. A **20**, 69 (1974)
7. M.J.G. Veltman, Conf. Proc. C **7507281**, 265 (1975)
8. B.P. Abbott et al., LIGO scientific and virgo collaborations. Phys. Rev. Lett. **116**(6), 061102 (2016), arXiv:1602.03837 [gr-qc]
9. K. Akiyama et al., Event horizon telescope. Astrophys. J. Lett. **875**, L1 (2019). ([arXiv:1906.11238 [astro-ph.GA]])
10. I.L. Buchbinder, S.D. Odintsov, I.L. Shapiro, *Effective Action in Quantum Gravity* (IOP, Bristol, UK, 1992), p.413
11. A.G. Riess et al., Supernova search team. Astron. J. **116**, 1009 (1998). ([astro-ph/9805201])
12. S.M. Carroll, Living Rev. Rel. **4**, 1 (2001). ([arXiv:astro-ph/0004075])
13. D.M. Ghilencea, JHEP **1903**, 049 (2019). ([arXiv:1812.08613 [hep-th]])
14. D.M. Ghilencea, Phys. Rev. D **101**(4), 045010 (2020), arXiv:1904.06596 [hep-th]
15. P.C. West, *Introduction to Supersymmetry and Supergravity* (World Scientific, Singapore, Singapore, 1990), p.425
16. K.S. Stelle, Phys. Rev. D **16**, 953 (1977)
17. K.S. Stelle, Gen. Rel. Grav. **9**, 353 (1978)
18. A. A. Starobinsky, Phys. Lett. **91B**, 99 (1980) [Adv. Ser. Astrophys. Cosmol. **3**, 130 (1987)]
19. S.W. Hawking, T. Hertog, Phys. Rev. D **65**, 103515 (2002). [hep-th/0107088]
20. R.P. Woodard, Lect. Notes Phys. **720**, 403 (2007). ([astro-ph/0601672])
21. I. Antoniadis, E. Mottola, Phys. Rev. D **45**, 2013 (1992)
22. A.V. Smilga, Nucl. Phys. B **706**, 598 (2005). ([hep-th/0407231])
23. M. Fontanini, M. Trodden, Phys. Rev. D **83**, 103518 (2011), arXiv:1102.4357 [gr-qc]
24. A. Salvio, A. Strumia, JHEP **06**, 080 (2014). ([arXiv:1403.4226 [hep-ph]])
25. L. Alvarez-Gaume, A. Kehagias, C. Kounnas, D. Lüst, A. Riotto, Fortsch. Phys. **64**(2–3), 176 (2016), arXiv:1505.07657 [hep-th]
26. A. Salvio, Front. Phys. **6**, 77 (2018). ([arXiv:1804.09944 [hep-th]])
27. A. Salvio, A. Strumia, H. Veermäe, Eur. Phys. J. C **78**(10), 842 (2018), arXiv:1808.07883 [hep-th]
28. A. Salvio, Phys. Rev. D **99**(10), 103507 (2019), arXiv:1902.09557 [gr-qc]
29. A. Salvio, Eur. Phys. J. C **79**(9), 750 (2019), arXiv:1907.00983 [hep-ph]
30. F.d.O. Salles, I.L. Shapiro, Phys. Rev. D **89**(8), 084054 (2014); Erratum: [Phys. Rev. D **90**, no. 12, 129903 (2014)], arXiv:1401.4583 [hep-th]

© The Author(s), under exclusive license to Springer Nature Switzerland AG 2023
A. Petrov et al., *Introduction to Modified Gravity*,
SpringerBriefs in Physics, https://doi.org/10.1007/978-3-031-46634-2

31. P.M. Lavrov, I.L. Shapiro, Phys. Rev. D **100**, 026018 (2019). ([arXiv:1902.04687 [hep-th]])
32. S.M. Carroll, V. Duvvuri, M. Trodden, M.S. Turner, Phys. Rev. D **70**, 043528 (2004) [astro-ph/0306438]
33. M.J. Reboucas, J. Santos, Phys. Rev. D **80**, 063009 (2009). ([arXiv:0906.5354 [astro-ph.CO]])
34. T. Harko, F.S.N. Lobo, *Extensions of f(R) Gravity: Curvature-Matter Couplings and Hybrid Metric-Palatini Theory* (Cambridge University Press, 2019)
35. S. Nojiri, S.D. Odintsov, Phys. Lett. B **735**, 376 (2014). ([arXiv:1405.2439 [gr-qc]])
36. G.J. Olmo, D. Rubiera-Garcia, Phys. Rev. D **86**, 044014 (2012), arXiv:1207.6004 [gr-qc]
37. F.S. Gama, J.R. Nascimento, A.Y. Petrov, P.J. Porfirio, A.F. Santos, Phys. Rev. D **96**(6), 064020 (2017), arXiv:1707.03440 [hep-th]
38. G.J. Olmo, D. Rubiera-Garcia, Int. J. Mod. Phys. D **21**, 1250067 (2012). ([arXiv:1207.4303 [gr-qc]])
39. F.S.N. Lobo, J. Martinez-Asencio, G.J. Olmo, D. Rubiera-Garcia, Phys. Lett. B **731**, 163 (2014). ([arXiv:1311.5712 [hep-th]])
40. G. Cognola, S. Zerbini, J. Phys. A **45**, 374014 (2012), arXiv:1203.5032 [gr-qc]
41. J.R. Nascimento, A.Y. Petrov, P.J. Porfirio, R.N. da Silva, Eur. Phys. J. C **83**, 331 (2023). ([arXiv:2302.09935 [gr-qc]])
42. S.D. Odintsov, V.K. Oikonomou, S. Banerjee, Nucl. Phys. B **938**, 935–956 (2019). ([arXiv:1807.00335 [gr-qc]])
43. L. Amendola, L. Giani, G. Laverda, Phys. Lett. B **811**, 135923 (2020), arXiv:2006.04209 [astro-ph.CO]
44. D. Lovelock, J. Math. Phys. **12**, 498 (1971)
45. N. Deruelle, L. Farina-Busto, Phys. Rev. D **41**, 3696 (1990)
46. C.C. Briggs, gr-qc/9808050
47. L. Randall, R. Sundrum, Phys. Rev. Lett. **83**, 3370 (1999). ([hep-ph/9905221])
48. L. Randall, R. Sundrum, Phys. Rev. Lett. **83**, 4690 (1999). ([hep-th/9906064])
49. D. Bazeia, A. Lobao, L. Losano, R. Menezes, A.Y. Petrov, Phys. Rev. D **92**(6), 064010 (2015), arXiv:1502.02564 [hep-th]
50. V.I. Afonso, D. Bazeia, R. Menezes, A.Y. Petrov, Phys. Lett. B **658**, 71 (2007). ([arXiv:0710.3790 [hep-th]])
51. D. Bazeia, A.S. Lobão Jr., R. Menezes, A.Y. Petrov, A.J. da Silva, Phys. Lett. B **729**, 127 (2014). ([arXiv:1311.6294 [hep-th]])
52. K. Hinterbichler, Rev. Mod. Phys. **84**, 671 (2012). ([arXiv:1105.3735 [hep-th]])
53. H. van Dam, M.J.G. Veltman, Nucl. Phys. B **22**, 397 (1970)
54. V.I. Zakharov, JETP Lett. **12**, 312 (1970)
55. D.G. Boulware, S. Deser, Phys. Rev. D **6**, 3368 (1972)
56. S.F. Hassan, R.A. Rosen, Phys. Rev. Lett. **108**, 041101 (2012), arXiv:1106.3344 [hep-th]
57. V.A. Kostelecký, R. Potting, Phys. Rev. D **104**, 104046 (2021), arXiv:2108.04213 [gr-qc]
58. C. de Rham, Living Rev. Rel. **17**, 7 (2014). ([arXiv:1401.4173 [hep-th]])
59. A.S. Goldhaber, M.M. Nieto, Rev. Mod. Phys. **82**, 939 (2010). ([arXiv:0809.1003 [hep-ph]])
60. C.A.S. Almeida, W.T. Cruz, R.V. Maluf, A.Y. Petrov, P.J. Porfirio, Eur. Phys. J. C **82**, 715 (2022). ([arXiv:2202.01154 [hep-th]])
61. G.R. Dvali, G. Gabadadze, M. Porrati, Phys. Lett. B **485**, 208 (2000). ([arXiv:hep-th/0005016 [hep-th]])
62. P. Hao, D. Stojkovic, Phys. Rev. D **90**(2), 024002 (2014), arXiv:1404.7145 [gr-qc]
63. E.A. Bergshoeff, O. Hohm, P.K. Townsend, Phys. Rev. Lett. **102**, 201301 (2009), arXiv:0901.1766 [hep-th]
64. E.A. Bergshoeff, O. Hohm, P.K. Townsend, Phys. Rev. D **79**, 124042 (2009), arXiv:0905.1259 [hep-th]
65. B. Ratra, P.J.E. Peebles, Phys. Rev. D **37**, 3406 (1988)
66. S. Tsujikawa, Class. Quant. Grav. **30**, 214003 (2013), arXiv:1304.1961 [gr-qc]
67. S. Deser, R. Jackiw, S. Templeton, Ann. Phys. **140**, 372 (1982) [Annals Phys. **281**, 409 (2000)] Erratum: [Annals Phys. **185**, 406 (1988)]
68. R. Jackiw, S.Y. Pi, Phys. Rev. D **68**, 104012 (2003) [gr-qc/0308071]

69. S.M. Carroll, G.B. Field, R. Jackiw, Phys. Rev. D **41**, 1231 (1990)
70. L. Alvarez-Gaume, E. Witten, Nucl. Phys. B **234**, 269 (1984)
71. D. Grumiller, N. Yunes, Phys. Rev. D **77**, 044015 (2008), arXiv:0711.1868 [gr-qc]
72. M. Satoh, S. Kanno, J. Soda, Phys. Rev. D **77**, 023526 (2008), arXiv:0706.3585 [astro-ph]
73. P.J. Porfírio, J.B. Fonseca-Neto, J.R. Nascimento, A.Y. Petrov, J. Ricardo, A.F. Santos, Phys. Rev. D **94**(4), 044044 (2016), arXiv:1606.00743 [hep-th]
74. P.J. Porfírio, J.B. Fonseca-Neto, J.R. Nascimento, A.Y. Petrov, Phys. Rev. D **94**(10), 104057 (2016), arXiv:1610.01539 [hep-th]
75. S. Alexander, N. Yunes, Phys. Rept. **480**, 1 (2009). ([arXiv:0907.2562 [hep-th]])
76. N. Yunes, F. Pretorius, Phys. Rev. D **79**, 084043 (2009), arXiv:0902.4669 [gr-qc]
77. A.F. Ferrari, M. Gomes, J.R. Nascimento, E. Passos, A.Y. Petrov, A.J. da Silva, Phys. Lett. B **652**, 174 (2007). ([hep-th/0609222])
78. R. Jackiw, V.A. Kostelecky, Phys. Rev. Lett. **82**, 3572 (1999). ([hep-ph/9901358])
79. S.L. Adler, Phys. Rev. **177**, 2426 (1969); J.S. Bell, R. Jackiw, Nuovo Cim. A **60**, 47 (1969)
80. R. Jackiw, Int. J. Mod. Phys. B **14**, 2011 (2000). ([hep-th/9903044])
81. T. Mariz, J.R. Nascimento, E. Passos, R.F. Ribeiro, Phys. Rev. D **70**, 024014 (2004) [hep-th/0403205]
82. J.F. Assunção, T. Mariz, J.R. Nascimento, A.Y. Petrov, JHEP **1808**, 072 (2018). ([arXiv:1805.11049 [hep-th]])
83. T. Mariz, J.R. Nascimento, A.Y. Petrov, L.Y. Santos, A.J. da Silva, Phys. Lett. B **661**, 312 (2008). ([arXiv:0708.3348 [hep-th]])
84. M. Gomes, T. Mariz, J.R. Nascimento, E. Passos, A.Y. Petrov, A.J. da Silva, Phys. Rev. D **78**, 025029 (2008) arXiv:0805.4409 [hep-th]
85. J.M. Chung, Phys. Rev. D **60**, 127901 (1999) [hep-th/9904037]
86. B. Altschul, Phys. Rev. D **99**(12), 125009 (2019), arXiv:1903.10100 [hep-th]
87. C. Brans, R.H. Dicke, Phys. Rev. **124**, 925 (1961)
88. S. Sen, A.A. Sen, Phys. Rev. D **63**, 124006 (2001) [gr-qc/0010092]
89. J.A. Agudelo, J.R. Nascimento, A.Y. Petrov, P.J. Porfírio, A.F. Santos, Phys. Lett. B **762**, 96 (2016). ([arXiv:1603.07582 [hep-th]])
90. S.W. Hawking, Commun. Math. Phys. **25**, 167 (1972)
91. G.W. Horndeski, Int. J. Theor. Phys. **10**, 363 (1974)
92. D. Bazeia, L. Losano, R. Menezes, J.C.R.E. Oliveira, Eur. Phys. J. C **51**, 953 (2007). ([hep-th/0702052])
93. A. Nicolis, R. Rattazzi, E. Trincherini, Phys. Rev. D **79**, 064036 (2009), arXiv:0811.2197 [hep-th]
94. K. Hinterbichler, M. Trodden, D. Wesley, Phys. Rev. D **82**, 124018 (2010), arXiv:1008.1305 [hep-th]
95. L. Heisenberg, C.F. Steinwachs, JCAP **2001**(01), 014 (2020), arXiv:1909.04662 [hep-th]
96. C. Deffayet, D.A. Steer, Class. Quant. Grav. **30**, 214006 (2013), arXiv:1307.2450 [hep-th]
97. S. Appleby, E.V. Linder, JCAP **1203**, 043 (2012). ([arXiv:1112.1981 [astro-ph.CO]])
98. C. Charmousis, E.J. Copeland, A. Padilla, P.M. Saffin, Phys. Rev. Lett. **108**, 051101 (2012), arXiv:1106.2000 [hep-th]
99. A.A. Starobinsky, S.V. Sushkov, M.S. Volkov, JCAP **1606**, 007 (2016). ([arXiv:1604.06085 [hep-th]])
100. M. Rinaldi, Phys. Rev. D **86**, 084048 (2012), arXiv:1208.0103 [gr-qc]
101. A. Anabalon, A. Cisterna, J. Oliva, Phys. Rev. D **89**, 084050 (2014), arXiv:1312.3597 [gr-qc]
102. A. Cisterna, C. Erices, Phys. Rev. D **89**, 084038 (2014), arXiv:1401.4479 [gr-qc]
103. E. Babichev, C. Charmousis, A. Cisterna, M. Hassaine, arXiv:2004.00597 [hep-th]
104. T.L. Curtright, D.B. Fairlie, arXiv:1212.6972 [hep-th]
105. J.R. Nascimento, A.Y. Petrov, P. Porfírio, A.F. Santos, Phys. Rev. D **102**, 104064 (2020), arXiv:2009.13242 [gr-qc]
106. D. Colladay, V.A. Kostelecky, Phys. Rev. D **55**, 6760 (1997). ([hep-ph/9703464])
107. D. Colladay, V.A. Kostelecky, Phys. Rev. D **58**, 116002 (1998) [hep-ph/9809521]
108. V.A. Kostelecky, Phys. Rev. D **69**, 105009 (2004) [hep-th/0312310]

109. V.A. Kostelecký, Z. Li, Phys. Rev. D **103**, 024059 (2021). ([arXiv:2008.12206 [gr-qc]])
110. G. de Berredo-Peixoto, I.L. Shapiro, Phys. Lett. B **642**, 153 (2006). ([hep-th/0607109])
111. V.A. Kostelecky, M. Mewes, Phys. Lett. B **779**, 136 (2018). ([arXiv:1712.10268 [gr-qc]])
112. S.M. Carroll, H. Tam, Phys. Rev. D **78**, 044047 (2008). ([arXiv:0802.0521 [hep-ph]])
113. R. Bluhm, V.A. Kostelecky, Phys. Rev. D **71**, 065008 (2005). ([hep-th/0412320])
114. O. Bertolami, J. Paramos, Phys. Rev. D **72**, 044001 (2005). ([hep-th/0504215])
115. J. Beltran Jimenez, A.L. Maroto, JCAP **0902**, 025 (2009), arXiv:0811.0784 [astro-ph]
116. J. Beltran Jimenez, A.L. Maroto, Phys. Rev. D **80**, 063512 (2009), arXiv:0905.1245 [astro-ph.CO]
117. V.A. Kostelecky, S. Samuel, Phys. Rev. D **39**, 683 (1989)
118. T. Jacobson, D. Mattingly, Phys. Rev. D **64**, 024028 (2001). ([gr-qc/0007031])
119. C. Eling, T. Jacobson, Phys. Rev. D **69**, 064005 (2004). ([gr-qc/0310044])
120. A. Paliathanasis, G. Papagiannopoulos, S. Basilakos, J.D. Barrow, Eur. Phys. J. C **79**(8), 723 (2019), arXiv:1906.03872 [gr-qc]
121. A. Paliathanasis, Phys. Rev. D **101**(6), 064008 (2020), arXiv:2001.02016 [gr-qc]
122. X. Meng, X.L. Du, Phys. Lett. B **710**, 493 (2012)
123. T. Jacobson, PoS QG -**PH**, 020 (2007), arXiv:0801.1547 [gr-qc]
124. D.J. Gross, A. Neveu, Phys. Rev. D **10**, 3235 (1974)
125. J.L. Chkareuli, J. Jejelava, Z. Kepuladze, Bled Workshops Phys. **19**(2), 74 (2018), arXiv:1811.09578 [physics.gen-ph]
126. R.V. Maluf, J.C.S. Neves, Phys. Rev. D **103**(4), 044002 (2021), arXiv:2011.12841 [gr-qc]
127. D. Capelo, J. Paramos, Phys. Rev. D **91**(10) 104007 (2015), arXiv:1501.07685 [gr-qc]
128. R.V. Maluf, J.C.S. Neves, JCAP **10**, 038 (2021). ([arXiv:2105.08659 [gr-qc]])
129. A.F. Santos, A.Y. Petrov, W.D.R. Jesus, J.R. Nascimento, Mod. Phys. Lett. A **30**(2), 1550011 (2015), arXiv:1407.5985 [hep-th]
130. W.D.R. Jesus, A.F. Santos, Int. J. Mod. Phys. A **35**, 2050050 (2020). ([arXiv:2003.13364 [gr-qc]])
131. M.D. Seifert, Phys. Rev. D **81**, 065010 (2010), arXiv:0909.3118 [hep-ph]
132. R.V. Maluf, C.A.S. Almeida, R. Casana, M.M. Ferreira, Jr., Phys. Rev. D **90**(2), 025007 (2014), arXiv:1402.3554 [hep-th]
133. J.R. Nascimento, A.Y. Petrov, A.R. Vieira, Galaxies **9**, 32 (2021), arXiv:2104.01651 [gr-qc]
134. C. Hernaski, H. Belich, Phys. Rev. D **89**, 104027 (2014), arXiv:1409.5742 [hep-th]
135. M. Mewes, Phys. Rev. D **99**, 104062 (2019), arXiv:1905.00409 [gr-qc]
136. Y. Bonder, Phys. Rev. D **91**(12), 125002 (2015), arXiv:1504.03636 [gr-qc]
137. R.C. Myers, M. Pospelov, Phys. Rev. Lett. **90**, 211601 (2003), arXiv:hep-ph/0301124 [hep-ph]
138. J.F. Assunção, T. Mariz, J.R. Nascimento, A.Y. Petrov, Phys. Rev. D **100**, 085009 (2019), arXiv:1902.10592 [hep-th]
139. P. Horava, Phys. Rev. D **79**, 084008 (2009), arXiv:0901.3775 [hep-th]
140. E.M. Lifshitz, Zh. Eksp. Teor. Fiz. 11, 255 & 269 (1941)
141. R.L. Arnowitt, S. Deser, C.W. Misner, Gen. Rel. Grav. **40**, 1997 (2008). ([gr-qc/0405109])
142. G. Giribet, D.L. Nacir, F.D. Mazzitelli, JHEP **1009**, 009 (2010), arXiv:1006.2870 [hep-th]
143. D.L. Lopez Nacir, F.D. Mazzitelli, L.G. Trombetta, Phys. Rev. D **85**, 024051 (2012), arXiv:1111.1662 [hep-th]
144. E. Kiritsis, G. Kofinas, Nucl. Phys. B **821**, 467 (2009), arXiv:0904.1334 [hep-th]
145. J.B. Fonseca-Neto, A.Y. Petrov, M.J. Reboucas, Phys. Lett. B **725**, 412 (2013), arXiv:1304.4675 [astro-ph.CO]
146. S. Mukohyama, Class. Quant. Grav. **27**, 223101 (2010), arXiv:1007.5199 [hep-th]
147. H. Lu, J. Mei, C. Pope, Phys. Rev. Lett. **103**, 091301 (2009), arXiv:0904.1595 [hep-th]
148. A. Kehagias, K. Sfetsos, Phys. Lett. B **678**, 123–126 (2009), arXiv:0905.0477 [hep-th]
149. D. Blas, O. Pujolas, S. Sibiryakov, JHEP **10**, 029 (2009), arXiv:0906.3046 [hep-th]
150. C. Charmousis, G. Niz, A. Padilla, P.M. Saffin, JHEP **08**, 070 (2009), arXiv:0905.2579 [hep-th]
151. D. Blas, O. Pujolas, S. Sibiryakov, JHEP **1104**, 018 (2011), arXiv:1007.3503 [hep-th]

152. S. Weinfurtner, T.P. Sotiriou, M. Visser, J. Phys. Conf. Ser. **222**, 012054 (2010), arXiv:1002.0308 [gr-qc]
153. A. Wang, Int. J. Mod. Phys. D **26**(7), 1730014 (2017), arXiv:1701.06087 [gr-qc]
154. G.V. Efimov, Commun. Math. Phys. **5**(1), 42 (1967)
155. T. Biswas, A. Mazumdar, W. Siegel, JCAP **03**, 009 (2006), arXiv:hep-th/0508194 [hep-th]
156. E.T. Tomboulis, hep-th/9702146
157. L. Modesto, Phys. Rev. D **86**, 044005 (2012), arXiv:1107.2403 [hep-th]
158. T. Biswas, T. Koivisto, A. Mazumdar, JCAP **1011**, 008 (2010), arXiv:1005.0590 [hep-th]
159. A. Bas Beneito, G. Calcagni, L. Rachwal, arXiv:2211.05606 [hep-th]
160. A. Smailagic, E. Spallucci, J. Phys. A **37**, 1–10 (2004), arXiv:hep-th/0406174 [hep-th]
161. F. Briscese, E.R. Bezerra de Mello, A.Y. Petrov, V.B. Bezerra, Phys. Rev. D **92**(10), 104026 (2015), arXiv:1508.02001 [gr-qc]
162. E.R. Bezerra de Mello, F.S. Gama, J.R. Nascimento, A.Y. Petrov, Phys. Rev. D **95**(2), 025028 (2017), arXiv:1611.09676 [hep-th]
163. F.S. Gama, J.R. Nascimento, A.Y. Petrov, P.J. Porfírio, Phys. Rev. D **96**(10), 105009 (2017), arXiv:1710.02043 [hep-th]
164. F. Briscese, L. Modesto, Phys. Rev. D **99**(10), 104043 (2019), arXiv:1803.08827 [gr-qc]
165. A.S. Koshelev, A. Tokareva, Phys. Rev. D **104**, 025016 (2021), arXiv:2103.01945 [hep-th]
166. F. Briscese, L. Modesto, Eur. Phys. J. C **81**, 730 (2021), arXiv:2103.00353 [hep-th]
167. T. Biswas, A.S. Koshelev, A. Mazumdar, S.Y. Vernov, JCAP **1208**, 024 (2012), arXiv:1206.6374 [astro-ph.CO]
168. I. Dimitrijevic, B. Dragovich, A. Koshelev, Z. Rakic, J. Stankovic, Phys. Lett. B **797**, 134848 (2019), arXiv:1906.07560 [gr-qc]
169. L. Modesto, L. Rachwal, Nucl. Phys. B **889**, 228 (2014), arXiv:1407.8036 [hep-th]
170. S. Deser, R.P. Woodard, Phys. Rev. Lett. **99**, 111301 (2007), arXiv:0706.2151 [astro-ph]
171. E. Elizalde, E.O. Pozdeeva, S.Y. Vernov, Phys. Rev. D **85**, 044002 (2012), arXiv:1110.5806 [astro-ph.CO]
172. Y. Li, L. Modesto, L. Rachwal, JHEP **12**, 173 (2015), arXiv:1506.08619 [hep-th]
173. J.R. Nascimento, A.Y. Petrov, P.J. Porfírio, Eur. Phys. J. C **81**, 815 (2021), arXiv:2102.01600 [gr-qc]
174. T. Biswas, A. Conroy, A.S. Koshelev, A. Mazumdar, Class. Quant. Grav. **31**, 015022 (2014) [erratum: Class. Quant. Grav. **31** (2014), 159501], arXiv:1308.2319 [hep-th]
175. Z. Zhao, L. Modesto, Eur. Phys. J. C **83**(6), 517 (2023), arXiv:2304.10318 [gr-qc]
176. M. Maggiore, M. Mancarella, Phys. Rev. D **90**(2), 023005 (2014), arXiv:1402.0448 [hep-th]
177. Y. Zhang, K. Koyama, M. Sasaki, G. Zhao, JHEP **03**, 039 (2016), arXiv:1601.03808 [hep-th]
178. K. Fernandes, A. Mitra, Phys. Rev. D **97**(10), 105003 (2018), arXiv:1710.09205 [gr-qc]
179. B. Mashhoon, *Nonlocal Gravity* (Oxford University Press, 2017)
180. A. Palatini, Rend. Circ. Mat, Palermo **43**, 203 (1919)
181. E.N. Saridakis et al. [CANTATA], arXiv:2105.12582 [gr-qc]
182. I.L. Shapiro, Phys. Rept. **357**, 113 (2002), arXiv:hep-th/0103093 [hep-th]
183. H. Weyl, Phys. Rev. **77**, 699–701 (1950)
184. O. Chandia, J. Zanelli, Phys. Rev. D **55**, 7580 (1997), arXiv:hep-th/9702025 [hep-th]
185. Y.N. Obukhov, A.J. Silenko, O.V. Teryaev, Phys. Rev. D **90**, 124068 (2014), arXiv:1410.6197 [hep-th]
186. J.R. Nascimento, A.Y. Petrov, P.J. Porfírio, Phys. Rev. D **105**(4), 044053 (2022), arXiv:2108.05705 [gr-qc]
187. A. Delhom, J. Nascimento, G.J. Olmo, A.Y. Petrov, P. Porfírio, Eur. Phys. J. C **81**, 287 (2021), arXiv:1911.11605 [hep-th]
188. A. Delhom, T. Mariz, J.R. Nascimento, G.J. Olmo, A.Y. Petrov, P.J. Porfírio, JCAP **07**, 018 (2022), arXiv:2202.11613 [hep-th]
189. A.O. Barvinsky, G.A. Vilkovisky, Phys. Rept. **119**, 1–74 (1985)
190. J.D. Bekenstein, Phys. Rev. D **70**, 083509 (2004) [erratum: Phys. Rev. D **71** (2005), 069901], arXiv:astro-ph/0403694 [astro-ph]
191. M.D. Seifert, Phys. Rev. D **81**, 065010 (2010), arXiv:0909.3118 [hep-ph]

192. S. Boudet, F. Bombacigno, G.J. Olmo, P.J. Porfirio, JCAP **05**, 032 (2022), arXiv:2203.04000 [gr-qc]
193. S. Alexander, N. Yunes, Phys. Rev. D **77**, 124040 (2008), arXiv:0804.1797 [gr-qc]
194. N. Bartolo, G. Orlando, JCAP **07**, 034 (2017), arXiv:1706.04627 [astro-ph.CO]
195. S. Melville, J. Noller, Phys. Rev. D **101**(2), 021502 (2020), arXiv:1904.05874 [astro-ph.CO]
196. P.S. Howe, K.S. Stelle, P.K. Townsend, Nucl. Phys. B **236**, 125 (1984)
197. P. Van Nieuwenhuizen, Phys. Rept. **68**, 189 (1981)
198. Z. Bern, L.J. Dixon, D. Dunbar, B. Julia, M. Perelstein, J. Rozowsky, D. Seminara, M. Trigiante, PoS **tmr2000**, 017 (2000), arXiv:hep-th/0012230 [hep-th]
199. Z. Bern, J.J. Carrasco, L.J. Dixon, H. Johansson, D.A. Kosower, R. Roiban, Phys. Rev. Lett. **98**, 161303161303 (2007). ([hep-th/0702112])
200. Z. Bern, J.J. Carrasco, L.J. Dixon, H. Johansson, R. Roiban, Phys. Rev. Lett. **103**, 081301 (2009), arXiv:0905.2326 [hep-th]
201. Z. Bern, J.J. Carrasco, W.M. Chen, A. Edison, H. Johansson, J. Parra-Martinez, R. Roiban, M. Zeng, Phys. Rev. D **98**(8), 086021 (2018), arXiv:1804.09311 [hep-th]
202. E.P. Verlinde, SciPost Phys. **2**(3), 016 (2017), arXiv:1611.02269 [hep-th]
203. M. Reuter, F. Saueressig, arXiv:0708.1317 [hep-th]

Index

© The Author(s), under exclusive license to Springer Nature Switzerland AG 2023
A. Petrov et al., *Introduction to Modified Gravity*,
SpringerBriefs in Physics, https://doi.org/10.1007/978-3-031-46634-2

Printed in the United States
by Baker & Taylor Publisher Services